the radio
amateur's
HANDBOOK

14th Revised Edition

the radio
amateur's
HANDBOOK

A. Frederick Collins

revised by
Robert Hertzberg, K4JBI

THOMAS Y. CROWELL, PUBLISHERS
New York *Established 1834*

Designed by Ruth Bornschlegel

Library of Congress Cataloging in Publication Data

Collins, Archie Frederick, date
   The radio amateur's handbook.

   Bibliography: p.
   Includes index.
   1. Radio—Amateurs' manuals.   I. Hertzberg, Robert
Edward, 1905– II. Title.
TK9956.C73   1979   621.3841   78-3303
ISBN 0-690-01772-3

79 80 81 82 83   10 9 8 7 6 5 4 3 2 1

Contents

1

Introduction to Amateur Radio

Greetings to all you newcomers, and especially to you former Citizens Band operators who are switching over to amateur radio in record numbers! You couldn't have picked a better time, because a whole new generation of exciting equipment is now evolving from solid-state technology, the operating range of very low-power transceivers is being increased enormously by the use of unmanned repeater stations, and hams are even getting into space communication by bouncing their signals off special amateur satellites and also the moon. Whether your interest is experimentation with electronic devices or chit-chat with other operators the world over, you'll find ham radio a fascinating, long-term hobby.

Just why are radio amateurs called *hams?* The word is so widely recognized that it appears in most major dictionaries, yet its background is vague. It probably originated in the heyday of wire-line telegraphy around the turn of the century, when operators tapped out messages in the American Morse Code on big brass keys. If they slapped the keys too hard and made too many false characters, other operators would refer to their hands rather derisively as *hams* in the sense that they were heavy and awkward. Consider a related expression, *ham actor,* which is popularly used to describe an inept performer. In the radio field, however, the word *ham* quickly lost its original uncomplimentary meaning as amateur operators became more skilled than many of their professional counterparts.

Today, hams everywhere refer to themselves almost exclusively as *hams* rather than as *radio amateurs*. Even the many female hams now on the air are hams and proud of it.

The No. 1 ham of all time must be an Italian student who was barely twenty years old in 1894. While vacationing in the Alps he found a copy of a magazine containing a description of some experiments made a few years previously by a young German professor named Heinrich Hertz. Hertz had set up two large loops of wire at opposite sides of his laboratory. To one he connected a pair of metal rods in a line, with the ends separated a fraction of an inch. To the other loop he added identical rods and to them connected a Ruhmkorff high-voltage induction coil, a device that later became the heart of the ignition systems of early automobiles. When the coil was energized by a large battery, sparks jumped across both gaps in unison. Since there was no physical connection between the loops, it was clear that electrical energy was bridging the space across the lab.

A budding scientist with a sharp, inquiring mind, the student reasoned that the range of transmission could be increased if the radiating elements were made larger and the receiver made more sensitive. Who was this bright lad? Guglielmo Marconi. Seven years later the name Marconi was a household word. Only fourteen years later, at the early age of thirty-four, he was a cowinner of the 1908 Nobel prize in physics. Not that he needed the money.

Marconi was more fortunate than other inventors of the period, or of any period for that matter: his family was rich and encouraged him in everything he did. His father was Italian, his mother Irish. He was completely bilingual and equally at home in both Italy and the British Isles.

In the public mind Marconi is generally regarded as the inventor of wireless telegraphy. This view is not fair either to him or to the many other experimenters who preceded him. Actually, what he did was to take some laboratory playthings out of the labs and into the field—literally into the field—and put them to work for practical communication through space.

His key contribution was the addition of an antenna system consisting of a long, high wire connected to one side of the transmitter and another wire buried in the ground and connected to the other side. To this day a grounded aerial is called simply a Marconi.

His demonstrations of two-way wireless telegraphy were dazzling successes. On December 12, 1901, Marconi and two assistants, in Newfoundland, received signals from England. For a startled world the age of international radio had begun.

Details of Marconi's equipment filled the technical press, and very quickly amateur radio became a popular hobby. Displaying great ingenuity, early hams assembled very successful stations out of odds and ends: milk bottles covered with tinfoil from cigarette boxes, to act as capacitors; plumbers' copper tubing formed into a helix, as transmitter tuning coils; spark coils from Model T Fords, for high voltage; No. 6 dry batteries from house bell systems, as a "power" source; oatmeal boxes wound with bell wire, as receiving tuning coils; bits of coal, as crystal detectors.

Spark transmitters, which were limited to code operation, were superseded rapidly after World War I by tube-type equipment, which provided voice as well as code capability. Using mostly homemade gear, hams were the first to demonstrate the great possibilities of short radio waves, previously considered absolutely useless. Hams love challenges and have a great record of winning them. And remember that this is all a labor of love, the fruits of which are shared by the whole fraternity. You'll enjoy being a ham!

2

Fundamentals
of Electricity

Electricity is what radio is all about. You undoubtedly learned a little about it in junior or senior high school, but didn't give it much thought later in life. Therefore, a quick refresher course on the subject is probably in order.

First, let us straighten out a matter ot terminology. The terms *electricity, electric current, current flow,* and *electronics* mean exactly the same thing: electrons in motion, performing useful work.

Scientists now pretty much agree that atoms, the building blocks of all matter, consist of a central core or nucleus of protons surrounded by a hoard of lighter particles called electrons. The protons and electrons have opposite electrical characteristics; arbitrarily, the protons are said to be positive and the electrons negative. These unlike charges attract each other with equal force, and in most materials they coexist in a state of balance. Like charges repel each other, so there is a latent tendency toward turbulence inside the atom. Being lighter than the protons, the electrons respond more readily to outside influences. Prior to World War II theoretical scientists had speculated for years about the potential energy represented by the atom's internal activity, and had come up with such astronomical figures that they were too frightened to believe them. By 1945, however, practical physicists had unlocked the secret of nuclear fission—that is, splitting the atom—and they produced the bomb that jarred the Japanese into surrender and ushered the world into the atomic era.

What makes electrons move? The application of other forms of energy: chemical, magnetic, thermal, and photovoltaic.

Batteries

With the advent of solid-state equipment, batteries have made a big comeback in the ham field. In various forms they are used widely in amplified microphones, test instruments, tape recorders, and especially in "walkie-talkies" (Fig. 2-1) and similar portable gear. All batteries produce direct current, DC, which flows evenly in one direction.

Batteries are chemical devices, consisting basically of two electrodes of dissimilar metals or other conductors surrounded by chemicals in either paste or liquid form, called the *electrolyte*. For example, the common flashlight battery uses a zinc can that acts as one electrode and also as the container for the paste electrolyte, in the center of which is stuck a bar or rod of carbon as the other electrode. It is called a "dry" battery because the container is sealed and normally keeps the paste inside. Batteries of this kind are satisfactory for short, intermittent service, but if left on too long their output drops badly. However, they do recuperate, up to a point, if allowed to rest.

A similar battery, known as the alkaline type, can furnish heavier current over longer periods.

A third kind of battery, called the mercury, is preferred for instruments that require a steady voltage for relatively long stretches. It does not fade away gradually, as the others do, but usually goes abruptly from normal voltage to nothing.

In producing current, the chemicals in dry batteries are eventually consumed. There is no practical way of rejuvenating them.

The lead-acid storage battery, familiar to anyone who has ever raised the hood of an automobile, does not really store electricity. It consists of two electrodes in an electrolyte, a liquid one this time, but it is not a self-contained, self-energizing battery like its dry cousins. The plates must first be formed, chemically, by a charging current from an outside source of electricity. After the latter is disconnected, the battery then delivers current of its own because of its chemical action. When it is near exhaustion, it can be recharged by another dose of outside current, and the process can be repeated thousands of times. The plates are not consumed, but simply change in character as current goes in and out. Some water is lost in the process, and should be replaced only with distilled water. Ordinary tap water contains impurities that can readily upset the chemical balance of the plates and the electrolyte and greatly shorten the life of the battery.

Small spillproof storage batteries that can substitute for flashlight size dry cells are known as Ni-Cads, for their electrodes of nickle and cadmium. Their output is high and steady, and they are easily maintained by chargers that accommodate two or four at a time.

2-1. Hand-held but no toy! Energized by a rechargeable NiCad battery pack, this complete 2-meter FM ham "station" from Wilson Electronics even includes a touch-tone pad for tieing in to land-line telephones by way of powerful repeater stations. Short antenna at the top of the case is called a "rubber duck."

Generators

One of the great discoveries of the nineteenth century was the relationship between electricity and magnetism: *each can produce the other*. The combination of a magnet and a coil of wire can generate a current in the latter if either element is moved in relation to the other. This is the basic design of all generators that furnish electric power to homes and industry.

Current passing through a coil of wire can create a magnetic field in and around it, and this field can induce a current in a second coil near it. This combination constitutes a transformer, used in a variety of forms in every radio transmitter and receiver ever built.

Low-Energy Sources

A union of two dissimilar metals, called a *thermocouple*, produces small currents of electricity when heated. The effect is of practical value only in thermostats for temperature control of heating systems, some kitchen appliances, and in certain frequency-control ovens used in radio equipment.

A material that produces electricity when light shines on it is called *photovoltaic*. This effect is of widespread value in camera exposure meters and, increasingly, in solar cells that produce enough juice in sunny areas to keep unattended storage batteries in charged condition. Hams are experimenting with these cells for use in remote short-wave repeater stations, and are reporting good results.

Voltage

How do we make electricity flow through a wire? To force water through a pipe we must exert pressure on it. The pressure can come from height—water flows from a high tank through a pipe to a low tank—or from a pump.

The same is true of electricity. Suppose we connect an area of negative charge to an area of positive charge by closing the switch in the wire between them (Fig. 2-2). Electrons will flow from the negative side to the positive side. It's as though one side were higher than the other. The force causing this flow is called the potential difference between the two sides.

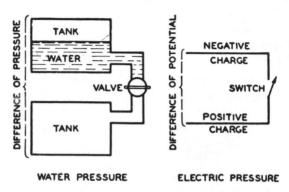

WATER PRESSURE   ELECTRIC PRESSURE

2-2. Electric potential (voltage) can be compared with water pressure.

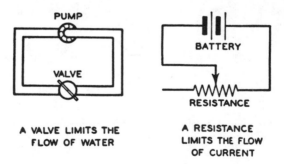

A VALVE LIMITS THE   A RESISTANCE
FLOW OF WATER   LIMITS THE FLOW
OF CURRENT

2-3. Electric resistance can be compared with a water valve.

We could use a pump to maintain pressure on water in a loop of pipe so that the water would continue to flow around the loop. There are electricity "pumps," too—batteries and generators. They maintain a potential difference between their positive and negative terminals so that there is a continuous electric pressure to push electrons around the circuit (Fig. 2-3).

Electric potential is expressed as *voltage* (the symbol is E). The number of volts tells how much electric pressure there is to force a current to flow. The instrument that measures potential is a voltmeter. It is connected across the wires of a circuit (Fig. 2-4).

Amperage

We often need to know the rate of flow, or *amperage*, in a circuit. One ampere is the amount of current that flows for one second. The symbol for current is I, and the instrument that measures current flow is an ammeter. An ammeter is always connected in series with the wires of a circuit (Fig. 2-4) so that all the current flows through it.

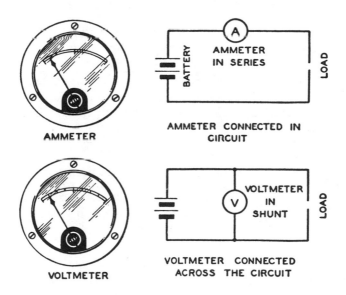

AMMETER

AMMETER CONNECTED IN CIRCUIT

VOLTMETER

VOLTMETER CONNECTED ACROSS THE CIRCUIT

2-4. How the ammeter and the voltmeter are used.

Resistance

Water flows more easily through a big, smooth pipe than through a small, rough one. The friction between the water and the pipe retards the flow.

Much the same is true of electric flow. A current flows more easily through a big wire than through a thin one. The kind of material that the wire is made of matters, too, for in some atoms the electrons can be nudged along more readily than in others. This opposition to the flow of electricity is called *resistance* (the symbol is R), and the unit of resistance is the *ohm*. On actual resistors, this is usually represented by the Greek letter Ω, for omega, the last letter of the Greek alphabet.

Materials are classified according to their resistance:

Conductors have very little resistance. They are metals; copper, aluminum, and silver are the best.

Resistors have enough resistance to limit the amount of current that can pass through them. They are made of carbon, of wire, or of metallic alloys.

Insulators have so much resistance that they permit practically no current to pass. They are substances like glass, plastics, ceramics, paper, cotton, and silk.

Semiconductors are special materials that generally have higher resistance than most conductors and lower resistance than most insulators. They are treated in detail in Chapter 4.

Ohm's Law

You can see that there should be a connection between voltage and the current it can force through a wire of a certain resistance. There is a mathematical relationship, Ohm's law, which says that the voltage across a wire equals the amount of current flowing in the wire (amperes) multiplied by the resistance of the wire (ohms). In symbols, this is written:

$$E = IR$$

So if you know the current and resistance, you can compute the voltage. If you know the voltage and resistance, you can compute the current, for

$$I = \frac{E}{R}$$

Or, if you know the current and the voltage, you can compute the resistance:

$$R = \frac{E}{I}$$

Power

Electricity is energy, the ability to perform work. When a resistance blocks the flow of some of this energy, it cannot destroy the blocked part but must get rid of it, dissipate it, in another form. A resistance converts electric energy into heat. The rate at which a resistance dissipates electric energy is measured in watts and is found by multiplying the current by itself and then by resistance. In symbols, this is written:

$$P = I^2 R$$

(I^2 means I squared, or I times I.)

If both the voltage and the current are known, the power in watts is simply:

$$P = EI$$

In direct-current circuits, Ohm's law is simple arithmetic. In alternating-current applications, it has a different form.

Capacitance

Metal plates or other conductive materials placed close together but kept from direct contact by an insulating medium such as air, paper, glass, or mica, constitute a *capacitor,* sometimes called a *condenser.* The insulating medium is the *dielectric.* A capacitor has the ability to store electric charges. This is called *capacitance,* and is measured in *farads* (f).

The charging action of capacitors is not very well understood, any more so than the actions of many other electronic devices. However, we can benefit by observing a simple experiment with a capacitor, a battery, a meter, and a switch, as shown in Fig. 2-5. Start at (A), with the switch in its off position; the circuit is open, completely dead. Close the switch to the No. 1 contact, as in (B); this puts the battery, the capacitor, and the meter into a simple series hookup. Now, since the

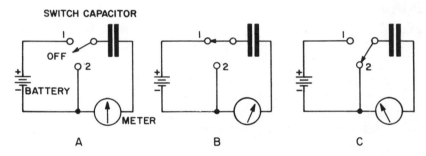

2-5. The charging and discharging actions of a capacitor are shown by movements of a meter in the circuit.

plates of the capacitor are separated by a good insulating material, the circuit should remain open and nothing should happen. But something very definite does happen: the meter needle kicks sharply upscale in one direction, which means without a shadow of a doubt that some electrons have gone through the entire circuit; then the needle drops almost as quickly to zero. Did this pulse of current dissipate itself in the resistance of the wires? To find out, first return the switch to its off position, thus again isolating all elements of the hookup. Then move it to the No. 2 contact, as in (C), putting the capacitor directly into the meter. The latter's needle again jerks violently, but this time in the *other* direction, and again comes to rest at zero.

It is fairly safe to assume that the first push of voltage from the battery causes a jamming of electrons in the highly resistive dielectric between the plates. A *few* electrons apparently nudge through; this accounts for the meter reading in what otherwise is an absolutely open circuit. Most of them, however, seem to remain on or between the plates, maintaining a static push against the dielectric much as if the latter were a coiled spring. This accumulation of electrons is considered the *charging* current.

Does removing the push of the battery voltage cause the charge in the capacitor to collapse? No! We remove the battery when we open the switch. Closing the latter puts the low-resistance meter across the capacitor. The charged dielectric uncoils, so to speak, and discharges the electrons back through the circuit the same way they entered. The slight dielectric leakage that occurs during charging recurs during discharging, to again account for the meter indication. In a fraction of a second the electrons in the circuit settle down to a quiescent state, and the capacitor and the meter both go dead, electrically.

Since the initial flick of current is only momentary, capacitors can be used in circuits to block off direct current where it isn't wanted. Capacitor function on AC is quite a different matter, and is taken up later in this chapter.

The pushing action of the electron charge on the dielectric of a capacitor has tangible physical effects. If one plate is thin and flexible, and if the applied voltage is high enough, variations of the latter make the plate vibrate. This is the basis of the electrostatic loudspeaker, which has been in existence since the 1920's.

It is also interesting to know that a capacitor that has been charged, and then removed completely from its circuit, can retain its charge for a long time: hours, days, or even weeks. However, it is not a really useful storage device in the sense that a storage battery is. The instant it starts discharging, the voltage starts to drop, and full discharge takes place in a very short time, usually a fraction of a second, sometimes more, depending on circuit conditions.

Inductance

In 1820 Hans Christian Oersted, a Dane, discovered that electricity and magnetism were related. He held a compass near a wire connected to a battery and noticed that the compass needle moved. It acted exactly as if it had been held near a magnet.

Moving electrons—electric currents—act just like permanent bar magnets. The current creates a field of magnetic force around the wire. The way this magnetic field is distributed in

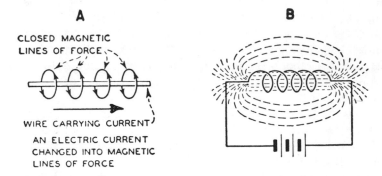

2-6. (A) Magnetic lines of force about a single conductor and (B) about a coil.

space is often indicated by drawing lines. Where the lines are closer together, the field is stronger and the magnetic force exerted is greater. You can see from Fig. 2-6 that a stronger field is created by shaping the current-carrying wire into a coil. The strength of the field is also increased by placing a piece of soft iron inside the coil. This iron bar is not a permanent magnet like the ones you buy in the dime store. It has no magnetism until electricity flows through the coil surrounding it, and it quickly loses its magnetism when the current is shut off.

A coil designed to create an electromagnetic field is technically named an *inductor*, but it is usually referred to simply as a coil. Its ability to create a magnetic field is called its *inductance* (symbol L) and is measured in henries (h).

Alternating Current

So far we have been talking of direct current. The kind that comes out of public utility powerhouses is quite different and is called *alternating current* because it changes constantly. Each complete change of alternating current (AC) is called a *cycle*, and consists of two flows of current, one in one direction in a circuit followed by the other in the opposite direction. The total number of cycles that occur in one second is called the *frequency* and is expressed by the term *hertz* in honor of Heinrich Hertz. (All basic electrical units such as volt, ampere, ohm, watt, farad, henry, etc., are named for early European and American experimenters.) The abbreviation is Hz, to distinguish it from the letter H, the unit of inductance. Thus, we refer to AC house power service as either 60 hertz or 60 Hz. The usual prefixes to indicate larger units apply, as with the other terms: *kilohertz* or *kHz* for 1,000 hertz, and *megahertz* or *MHz* for 1,000,000 hertz. Note that a capital M is used here for *mega*, because a small m represents the prefix *milli*, 1/1000.

The word hertz replaces an older term *cycles per second*, usually shortened to *cps*, found in most electronic literature published prior to about 1970.

You might compare direct current to water being pushed around a loop of pipe by a rotary pump (Fig. 2-7). The drops of water always flow in the same direction around the pipe loop. If we replaced the rotary pump with a piston pump, however, the water drops would move first one way, then the

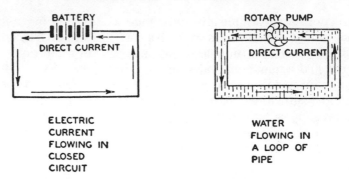

2-7. Electric current flow can be compared with flow of water in a pipe.

other way, following the back-and-forth motion of the piston (Fig. 2-8). Much the same sort of thing happens when an alternating-current generator pushes AC around an electric circuit.

Suppose we had a special kind of ammeter that could measure how much alternating current was flowing to a lamp at one particular instant, then how much current was flowing a fraction of a second later, and so on. We would find that at one instant there was no current at all—zero amperes. A brief fraction of a second later there would be a little current in one

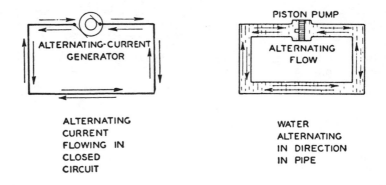

2-8. Water analogy for alternating current.

direction. A short time after that there would be still more current in that direction. As we continued our measurements, the
current would continue to increase in the same direction until
it reached a maximum value. Then it would begin to decrease
gradually to zero again. After the current had reached zero, it
would begin to increase in the opposite direction. It would
increase to a maximum amount in this direction, then decrease once more to zero. And then the current would start increasing in the first direction and the cycle would start all
over again.

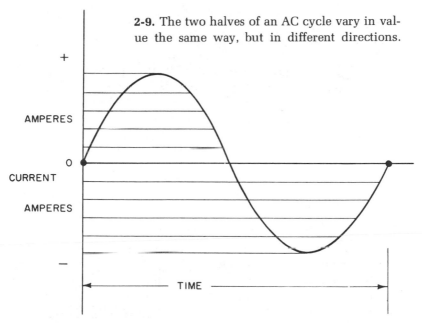

2-9. The two halves of an AC cycle vary in value the same way, but in different directions.

We could draw a graph of the value of the current, marking
off the time of measurement along the horizontal line of the
graph. The number of amperes at each instant would be indicated by the distance of our graph curve above or below the
horizontal time line—we'd make a mark above the line for
current flowing in one direction, below the line for current

flowing in the opposite direction. The graph would look like Fig. 2-9. If we measured AC *voltage* at successive instants of time, a graph of the results would be exactly the same as the graph of current.

This kind of graph looks strangely like a picture of a wave, doesn't it? If you drop a stone into a still pond, you make the water move up and down (Fig. 2-10). The first up-and-down motion, or wave, makes adjoining drops of water move up and down, and so on. The waves move rapidly out across the pond. Notice that the water itself does not move across the pond, but just moves up and down. It is the up-and-down *motion* that travels across the pond. Since motion is energy, a wave is traveling energy.

An alternating current of electricity is an electric wave. What about the electromagnetic field that is created by such an alternating current? If the current alternates, the field alternates, too. Its alternations follow the same pattern as that of the current. So now we have another wave, an electromagnetic wave. This is a radio wave and it carries electromagnetic energy just as a water wave carries mechanical energy.

Radio Waves

If we make a graph of a radio wave—that is, measure the way electromagnetic intensity changes with time—we get a picture that resembles the one made by alternating electricity. Remember that these graphs are not true images of the physical appearance of electric or radio waves—you can't see these waves, so actually they don't look like anything. The graphs merely illustrate the mathematics of wave behavior.

We can gather some useful information about waves from their graph-pictures. The height of the peak, or the depth of the trough, is the maximum *amplitude* of the wave. The distance between peaks (or troughs) is the *wavelength*. The change in the wave from one peak to the next (or one trough to the next) is called a *cycle* (heavy line in Fig. 2-10D). And

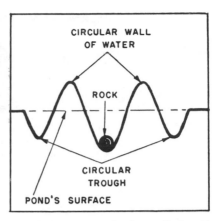

A. Dropping a rock into a still pool causes waves that move away from the splash in ever-widening circles.

B. Cross-sectional view of the first circular wall built up from the water displaced by the rock.

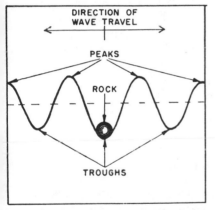

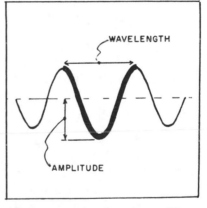

C. Cross-sectional view showing how a series of peaks and troughs are built up. Notice that each succeeding peak and trough is slightly lower and shallower.

D. Cross-sectional view illustrating the terms wavelength, amplitude, and cycle. The heavy line represents one cycle.

2-10. A study of wave motion in water.

the number of cycles that occur in one second—that is, the number of waves that pass a given point in one second—is the *frequency* of the wave.

If we multiply the frequency of a wave by its wavelength we get the speed with which the wave travels. We know that radio waves move at about the speed of light, 300,000,000 meters per second. So if either the frequency or wavelength of a wave is specified, we can compute the one that is not specified.

AC Values

How can we specify the amperage or voltage of an alternating current? The values are constantly changing. We have to settle on an *effective* value that depends on power dissipated in a resistance. The effective alternating current equals the direct current which dissipates just as much heat as the alternating current. This turns out to be .707 times the maximum, or peak, value of the alternating current (or voltage). Alternating-current ammeters and voltmeters read effective values directly. If an AC voltmeter indicates 70.7 volts, the voltage is varying between zero and 100 volts. But 70.7, the effective voltage, is the value that is used.

Resistance, Reactance, and Impedance

Alternating current encounters resistance when passing through a wire just as direct current does. At low frequencies—within the audio-frequency range—the AC resistance is for all practical purposes identical to the DC resistance.

As frequencies become higher, however, the current becomes concentrated in the outer surface of the wire instead of being distributed uniformly through its thickness. At very high frequencies practically all the current flows near the surface of a conductor; in fact, a hollow tube conducts such frequencies just as well as a solid wire. This means that the resistance of a particular piece of wire depends on the frequency of the current, if the frequency is high. The radio-frequency resistance, therefore, may be quite different from the audio-frequency or DC resistance.

Alternating current causes an even more marked change in the action of capacitors. Depending on its capacitance and the frequency of the current in the circuit, a capacitor that blocks DC completely can pass AC with very little hindrance. It may not be correct from the theoretical standpoint to say that it *conducts* the current, because after all the plates *are* insulated from each other. From the practical standpoint, however, capacitors pass AC over a very wide range of frequencies.

What apparently happens in a capacitor working on AC is that the rapid charging and discharging action, under the influence of the constantly varying current, keeps the dielectric in a constant state of agitation. This is virtually equivalent to a movement of electrons, and by definition a movement of electrons is electricity. The explanation is reasonable if we consider that a capacitor's resistance to AC (more correctly, its *reactance*; see below) goes down sharply as the frequency of the AC goes up. In other words, the greater the agitation the greater the internal electron movement.

A capacitor's opposition to AC is called *capacitive reactance* (symbol X_C) and is greater for small capacitances and low frequencies than for high capacitances and high frequencies.

Reactance is measured in ohms, like resistance, and it limits the flow of current, like resistance, but it is not the same as resistance. The difference is that a resistor dissipates electric energy, but a capacitor can only store electric energy, not dissipate it. This again is the theory. There is unavoidably some dissipation of energy in the dielectric which shows up unmistakably as heat. In fact, some capacitors get quite warm in normal operation.

An inductor also acts differently with AC than with DC. Here is why. Any current flowing through a coil creates an electromagnetic field around the coil. If the current changes, as AC does, the electromagnetic field also changes. Now, a changing electromagnetic field generates electricity in a wire. So AC in a coil generates electricity in the same wire. This *induced voltage* is entirely separate from the voltage that causes the original current to flow.

The induced voltage retards the flow of the current that causes it when that current is increasing. It aids the flow of the current when that current is decreasing. So a coil always opposes changes in current flowing through it. The net effect of a coil is to restrict the flow of alternating current. This restriction is called *inductive reactance* (symbol X_L). It increases with increasing inductance and increasing frequency.

Inductive reactance is also measured in ohms, but, like capacitive reactance, it is not the same as resistance—a coil only stores electric energy, and does not dissipate it. There can be, however, the normal loss of energy in the straight DC resistance of the wires.

The total resistance to AC by a capacitor or inductor is the combination of its internal ohmic resistance (that of its metal conductive elements) and its reactance. This is called *impedance*, to distinguish it from the DC term *resistance*. As the dielectric of a capacitor by its very nature has very high resistance, the impedance and reactance of most capacitors are virtually alike. However, inductors always contain wire in amounts varying from small to large, so their impedance figures are higher than either their resistance and reactance values individually. Impedance is always expressed in ohms. The mere use of the word impedance signifies AC operation of some sort.

Transformers

Suppose we place two coils near each other and send AC or any other varying current through one of them. The changing electromagnetic field of the first coil will induce electricity in the second coil. The voltage and amperage or the induced electricity will depend on the number of turns of wire in each coil. If the second coil, or secondary, has more turns than the first coil, or primary, the secondary voltage will be higher than the primary voltage. But the secondary amperage will be lower than the primary amperage. If the secondary has fewer turns, its voltage will be lower and its amperage higher.

Coils paired like this are called *transformers* (Fig. 2-11). There is a big one on a light pole near your house to "step down" the high voltage of the power line to a lower voltage for your house wiring. Smaller transformers are used in radio

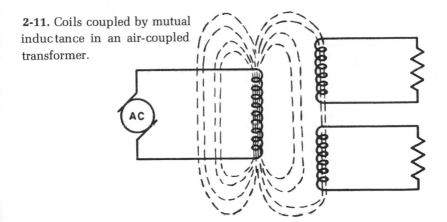

2-11. Coils coupled by mutual inductance in an air-coupled transformer.

equipment to step down or step up your 110-volt house power to the different voltage required for various parts of the equipment. The same principle of electromagnetic induction is also the basis for the operation of smaller transformers intended for radio frequencies.

Phase

When alternating current flows through a resistance, the voltage and amperage change their values together. They always increase or decrease in the same direction at the same time. A graph of one would eactly fit over a graph of the other (see Fig. 2-12).

This is not true of capacitors or coils. When the AC voltage applied to a capacitor is zero, the current flowing in the circuit is at its maximum. When the voltage is at its maximum, the current flowing is zero.

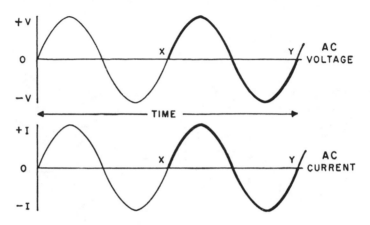

2-12. Sine waves of AC voltage and current.

You can see why. The voltage determines the amount of stored charge. As the voltage is first building up from zero, the capacitor contains no charge and no excess electrons and can therefore accept the greatest number of electrons—the greatest amount of current flow. At maximum voltage, the capacitor is fully charged and can accept no more electrons, so no current can flow into it.

The same sort of thing happens in a coil because of the way the induced voltage from its electromagnetic field opposes the applied voltage. The current is zero when the voltage is greatest, and vice versa.

In a capacitor, the current starts its cycle before the voltage and is said to *lead* the voltage. In a coil, the current starts its

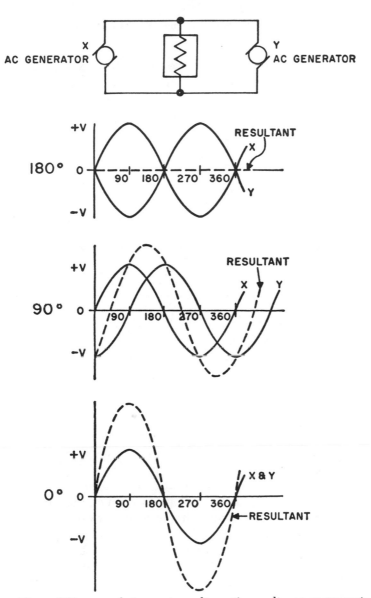

2-13. Phase differences between two alternating voltages or currents. Note that when two currents of equal value (first curves) are 180 degrees "out of phase" they buck each other and cancel out completely. At the other extreme (bottom curves), when they are in phase at zero degrees, they add to twice the value.

cycle after the voltage and is said to *lag* the voltage (see Fig. 2-13).

This difference in the timing of cycles is described in terms of *phase*. If current and voltage change together, as in a resistor, they are in phase. When one lags behind the other, they are out of phase.

The difference in phase is measured in fractions of a cycle, but it is not expressed as ½ cycle, ¼ cycle, and so on. Instead, a cycle is said to contain 360 degrees, and the phase difference is expressed in degrees, ½ cycle being 180 degrees, ¼ cycle 90 degrees, etc. (Fig. 2-14). If current increases while voltage decreases in the same direction, the phase difference is 90 degrees. If current increases and decreases in one direction while voltage does the same in the opposite direction, they are 180 degrees out of phase.

Phase difference is also used to describe the relationship between two or more voltages or two or more currents.

Circuits

Radio receivers and transmitters contain a great many resistors, capacitors, and coils connected together in circuits. Capacitors and coils are used to block currents of unwanted frequencies, since reactance depends on frequency. Resistors are used to reduce current or voltage. Transformers are used with AC to change current or voltage—increasing or decreasing one at the expense of the other.

Often you will need to calculate how much current is flowing through a circuit or how much voltage exists across a circuit. You may want to adjust current or voltage by adding or removing resistances or transformers. With DC, the calculations involve only resistance, since capacitors block DC completely. With AC, resistance, capacitive reactance, and inductive reactance must all be considered.

There are two main circuit arrangements. A *series* circuit has its elements connected one after another, so that the same

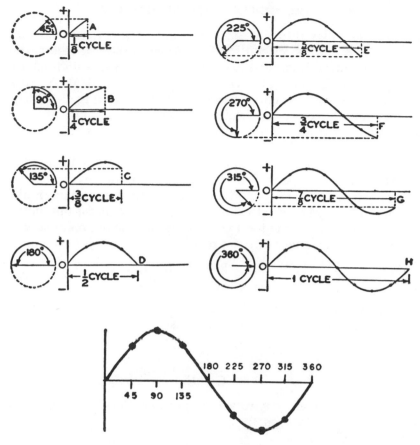

2-14. Evolution of an AC sine wave into electric degrees.

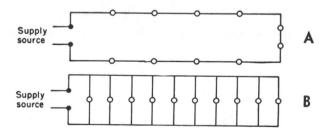

2-15. Simple series (A) and parallel (B) circuits.

current flows through each of them, in succession (Fig. 2-15A.). A *parallel* circuit has its elements arranged so that the same voltage is applied to each and the current is divided up, part flowing through each element (Fig. 2-15B). Both types may be combined into a *series-parallel* circuit (Fig. 2-16). For calculating current and voltage, a series-parallel circuit is divided into subcircuits, each of which is all series or all parallel.

The first step in analyzing a circuit is to find the total of the resistance or capacitance or inductance in the circuit. For resistors wired in series (Fig. 2-17), the total resistance is the sum of the individual values. The same is true for the inductance of coils, provided the coils are far enough apart so that their magnetic fields do not overlap.

$$R_{total} = R_1 + R_2 + R_3 \text{ etc.}$$
$$L_{total} = L_1 + L_2 + L_3 \text{ etc.}$$

Capacitors in series offer less total capacitance than any individual one would. The formula is:

$$C_{total} = \cfrac{1}{\dfrac{1}{C_1} + \dfrac{1}{C_2} + \dfrac{1}{C_3} \text{ etc.}}$$

If resistors are connected in parallel (Fig. 2-18), the total resistance is reduced:

$$R_{total} = \cfrac{1}{\dfrac{1}{R_1} + \dfrac{1}{R_2} + \dfrac{1}{R_3} \text{ etc.}}$$

Again, the same rule applies to the inductance of coils:

$$L_{total} = \cfrac{1}{\dfrac{1}{L_1} + \dfrac{1}{L_2} + \dfrac{1}{L_3} \text{ etc.}}$$

Capacitors in parallel, however, add their individual values:

$$C_{total} = C_1 + C_2 + C_3 \text{ etc.}$$

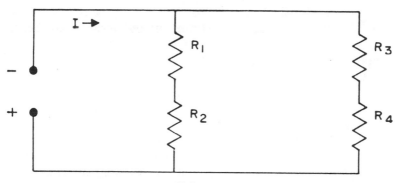

2-16. Resistances in series-parallel.

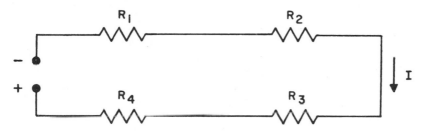

2-17. Resistances in series.

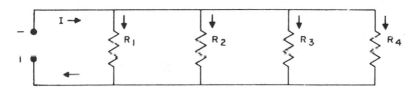

2-18. Resistances in parallel.

You can calculate voltage or current for a circuit containing resistors alone by simply applying Ohm's law to the total resistance:

$$E = IR_{total}, \qquad I = \frac{E}{R_{total}}, \qquad R_{total} = \frac{E}{I}$$

Ohm's law also works for alternating current circuits containing capacitors alone or coils alone, but you must remember to use the capacitive reactance ($X_C = 1/2\pi f C_{total}$) or inductive reactance ($X_L = 2\pi f L_{total}$) in the formula, not the capacitance or inductance directly.

$$E = IX_C, \qquad E = IX_L$$

Most AC circuits contain all three elements mixed together. All three affect the flow of current. The total opposition they offer (that is, the effect of capacitance, inductance, and resistance combined) is called *impedance* (symbol Z). Ohm's law applies to impedance as well as to resistance or reactance individually:

$$E = IZ, \quad I = \frac{E}{Z}, \qquad Z = \frac{E}{I}$$

The impedance of an AC circuit is not the simple sum of its resistance and reactance. For a circuit containing resistance, capacitance, and inductance in *series* (Fig. 2-19), the total impedance is given by this formula:

$$Z = \sqrt{R^2 + (X_L - X_C)^2}$$

that is, the square root of a sum made up of the square of the resistance plus the square of the difference between inductive reactance and capacitive reactance.

If a circuit contains elements in parallel, it is divided into its branches, and the current in each branch can be calculated by treating the branches as series circuits. You cannot, however, simply add up these individual branch currents to get the total current. Phase differences must be taken into account, and these can lead to complicated calculations on the engineering level. Fortunately, the problem rarely arises in amateur radio operation.

A mere understanding of the foregoing formulas can be a great help in maintenance and repair work on electronic equipment of all kinds. For example, consider ordinary resistors, which are used by the hundred in individual pieces of

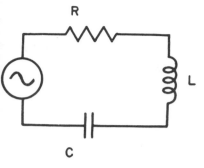

2-19. An AC circuit with resistance, inductive reactance, and capacitive reactance.

ham gear. Because their very circuit function is to control current by dissipating some of it in the form of heat, they often either change in value or burn out after long periods of service. It is very easy to check a resistor with the simple VOM described in Chapter 9. The problem is to find a replacement. Suppose the circuit diagram or parts list for the equipment identifies the faulty item as 500,000 ohms. You rummage through your "junk box" (which is an inevitable part of every ham shack) and can't find one of that value. Sort them according to resistance, and with a little elementary arithmetic take out enough of them to total 500,000 ohms, plus or minus 10 or 15 percent. Connect them in *series* like a bunch of firecrackers, install in the set, and you're back on the air. If you're lucky enough to find two one-megohm resistors among your treasures, connect them in *parallel,* and the effective resistance becomes 500,000 ohms.

The parallel method of buildup is more usual when a relatively low-value resistor is needed. Remember here that any resistor, regardless of value, connected across any other resistor produces a resistance lower than that of either. For example, suppose you need 50 ohms, but the smallest one you can find measures 300 on your VOM. Start adding others to it in parallel, at the same time observing on the VOM how the total resistance goes down. This is much less work than wrestling with the reciprocal fractions of the formula for resistors in parallel.

Small capacitors of the mica, paper, and ceramic types rarely need replacement, which is a good thing because they are not as easy to check as resistors. A dead, internal short-circuit will bang the VOM needle to the zero resistance mark, so you know the capacitor is useless. However, no reading does not necessarily mean it is okay, especially if it is of low value; an internal lead might simply be broken. With higher capacitance values, a momentary upswing of the meter needle is a pretty fair indication that the component is in normal condition. See the section entitled "Capacitance" earlier in this chapter for an explanation of this action.

Resonance

In some respects, capacitors and coils are like springs. They store electric energy, while springs store mechanical energy.

When you strike a spring, it vibrates, alternately storing and releasing mechanical energy. If you continue to strike it at just the right rate, or frequency, it will vibrate wildly. This very strong vibration at a preferred frequency is called *resonance*. If a car goes over a series of bumps at a certain speed, the front wheels sometimes bounce up and down very strongly. This is resonance involving the front-wheel springs. The bumps are coming at just the right frequency to excite resonance.

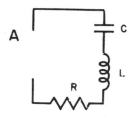

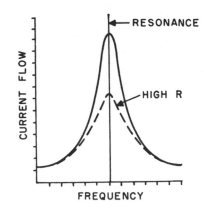

2-20. (A) Series resonant circuit and (B) graph of current flow vs. frequency in a series resonant circuit.

The electrical equivalent of vibration is called *oscillation*. Almost any circuit containing capacitance and inductance can be made to oscillate, at a frequency determined by the values of the two elements. When you *tune* a radio receiver or transmitter you change the value of either a capacitor or a coil, or both.

It makes a difference whether the inductance and capacitance are connected in series or parallel. With a series tuned circuit (Fig. 2-20), the current through the circuit will be high at the resonant frequency—the circuit behaves like a low resistance. With a parallel tuned circuit, the voltage across the circuit will be high at the resonant frequency—the circuit behaves like a high impedance (Fig. 2-21).

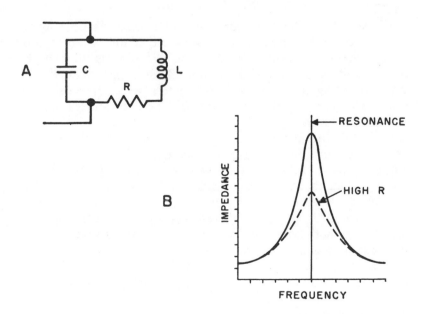

2-21.   (A) Parallel resonant circuit and (B) graph of impedance vs. frequency in a parallel resonant circuit.

The Electric "Ground"

In the early days of radio, battery operated receivers were constructed with front panels and interior chassis of Bakelite, hard rubber, and other good insulating materials. It was thought at the time that this insulation was necessary to prevent leakage of weak signals. The equipment was very sensitive to "hand capacitance"; that is, the tuning would change as the operator moved his hands from one dial to another.

To eliminate this effect, which made shortwave reception particularly difficult, engineers went to the other extreme. They did away with the insulated panels and subpanels and instead used sheet metal throughout. With the all-metal chassis connected to a water- or steam-pipe ground, hand capacitance disappeared, and tuning became comfortable and reliable.

The surface of the earth apparently acts as one large plate of a capacitor, the human body as one small plate. When an operator touches the grounded radio set, he connects himself directly to the ground and thus kills off his capacitance effect. With the older insulated sets, this action took place through the unshielded, unprotected circuit elements and naturally caused their resonances to shift.

In its full literal sense, an electric ground is an actual connection to earth. In some cases this is made simply by driving a metal pipe into the ground. In all urban and many suburban areas perfect ready-made grounds exist in the form of water pipes buried below the frost line.

The ground is an essential part of some transmitting antennas. It acts both as an artificial half of an aerial and as a reflecting surface to shoot the radio signals into space at various angles.

Any conductive body or surface that is large in comparison with associated equipment is considered a ground. Thus, the body of an automobile, although it is very effectively insulated from the actual ground by the rubber tires of the vehicle, is an excellent ground for a whip antenna mounted on a bumper or fender.

In many installations it is found that removing the actual ground wire from the receiver or transmitter makes no noticeable difference in operation. This is possible because a very low impedance path, at radio frequencies, is provided between the chassis and the grounded AC power line by the considerable capacitance effect between the primary and the various secondary windings of the power transformer. However, the ground wire should be retained to furnish a direct, low-resistance path to ground for possible static charges.

Because the chassis is grounded, directly or indirectly, all connections to it are called grounds. The chassis is the common return path for practically all the DC, AC, and RF circuits in a unit. It is remarkable that the currents in these circuits circulate without any mutual interference whatsoever.

3

Vacuum Tubes

Although home-entertainment equipment has gone completely solid-state in recent years, much ham gear still uses tubes. That's why you should know something about them.

Space or Emission Current

Visualize an electric light bulb, to which has been added a small metal plate close to but not touching the filament. A wire attached to the plate is brought out through the side of the glass. If we connect this wire to the positive pole of a battery, the negative pole to one side of a sensitive ammeter, and the other side of the ammeter to either leg of the filament, and then turn the lamp on, something interesting happens: The meter needle moves up-scale, indicating that current is flowing across the open space between the filament and the plate. How does it get across this gap?

When the tungsten filament is cold, the negative electrons that are part of its atomic structure are in a quiet state. When the bulb is turned on, current rushes through the wire and makes it white hot. The surface electrons on the wire become agitated and try to break away, but in an ordinary lamp they are restrained by the pulling action of the positively charged nucleus of the metal. However, in our special lamp, the positive attraction of the external battery is greater than that of the nucleus, so some electrons do break loose from the filament and slam into the plate to complete the simple series circuit

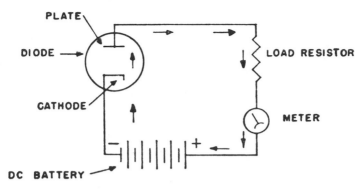

3-1. One-way current conduction in a diode. This diagram also shows the zigzag symbol for a resistor, the circle for a meter, and the alternate short and long lines for a battery. Of the latter, the long line is always the positive side, the short line the negative.

consisting of the lamp, the battery, and the ammeter. This flow of electrons is called *space* or *emission current* See Fig. 3-1.

The word "slam" here is not a mere figure of speech. In many high-power radio transmitting tubes the electrons hit the plate with such force and create so much friction that it can readily glow red in normal operation.

While any primary source of electrons in a vacuum tube is technically a *cathode*, this term in common practice usually indicates a small metal cylinder coated on the outside with chemicals that emit electrons copiously with rising temperature. Inside the cylinder and insulated from it is a spiral of wire heated by AC or DC passing through. Called the *heater*, this wire heats the cylinder by close radiation. It throws off very few electrons of its own because it glows only dull red and also because it is shielded by the cathode from the influence of the positively charged plate. See Fig. 3-2.

The term *filament*, by contrast, is generally understood to mean a bare wire burning at rather high temperature and pushing out electrons directly.

The cathode method of producing electron flow is widely favored because the relatively cool heater lasts a very long

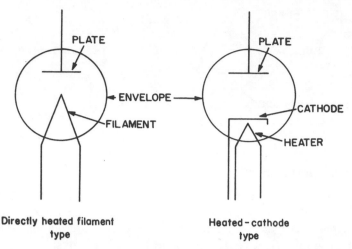

Directly heated filament type

Heated‑cathode type

3-2. Circuit symbols for the diode tube are in effect pictures of the internal construction.

time and can withstand accidental overloads that would quickly cause an incandescent filament to pop open. It is not at all unusual to hear of heater-type tubes that continue to work in perfectly normal fashion after twenty-five or more years of service.

The plate element is called the *anode*, in accordance with the standard practice of identifying the positive leg of a circuit or device. The *cathode* is always the negative leg.

The Diode

The basic electronic tube is the *diode*, consisting of two elements: the anode and a cathode in a glass bulb from which the air has been pumped out to form a vacuum. Around the turn of the century it greatly excited the early wireless experimenters who were searching frantically for a reliable rectifier (*detector*, they called it) of radio signals. It didn't take them very long to figure out that if a positive voltage on the plate of the tube attracted negative electrons from the cathode, a nega-

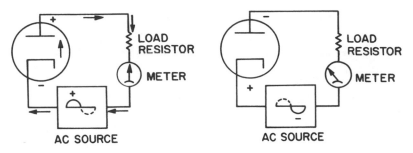

3-3. A diode is both a rectifier of AC and an effective electronic switch. As a switch, it is "closed" when the plate is positive and "open" when it is negative.

tive charge should repel them and thus in effect open the anode-to-cathode circuit completely. To prove this reasoning correct, they had only to flip the connections to the battery back and forth and watch the ammeter. Obviously, the next step was to try alternating current on the tube. See Fig. 3-3. As expected, it passed current only during the positive alternation and acted like a wide-open switch during the negative one. The diode was hailed as the perfect rectifier for changing AC into pulsating but nevertheless unidirectional current.

Perfect for some purposes, but unfortunately a flop for radio reception. It worked after a fashion when the positive alternations of the incoming impulses were strong enough to attract electrons, but the action was erratic and undependable because the tubes of the period were mostly handmade and imperfectly evacuated. They were soon replaced by very simple but far superior solid-state crystal detectors that cost literally only pennies yet provided clear reception over long distances. (See next chapter.)

The Triode

In 1906 Lee de Forest, an American inventor, added a third element, called the grid, to the diode and turned this former laboratory toy into the most far-reaching technical achievement

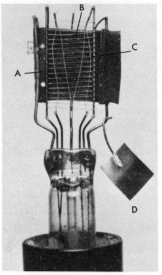

3-4. One side of the plate (A) of a basic triode of the filament type has been cut away here to show the open-mesh grid (B) and the central M-shaped filament (C). The square little plate (D) is a chemically coated "getter." After the tube is assembled in its envelope and evacuated, this "getter" is made to glow red hot by a current induced in it by an outside coil. The chemical flashes up and consumes any air or other gas still remaining on the elements of the tube.

of the twentieth century. Called the *triode,* the new tube not only could amplify electrical signals of all kinds to enormous values but could also generate easily adjustable alternating currents ranging in frequency from a few hertz per second to billions of hertz per second. It made possible the marvels we now take for granted: radio in all its forms, television in both black-and-white and color, talking movies, long-distance telephony, hearing aids, high-fidelity sound systems, to name just a few. How did a few turns of thin wire accomplish these miracles?

You already know that a diode consists of an anode and a cathode. De Forest put the grid between these elements, but closer to the latter. See Fig. 3-4. The wires of the grid are well separated and do not appreciably interfere with the movement of liberated electrons from the cathode. Now consider circuit A of Fig. 3-5. The plate circuit contains a battery, a meter, and a load resistor. Between the grid and the cathode, which is the *grid circuit,* there is a second battery or other source of DC, called the *bias supply* or the *grid bias.* For purposes of demonstration this is arranged with a switch so that the grid can be made either negative or positive in relation to the cathode; at the same time, the bias is adjustable in value.

With the grid switch open, the grid is isolated and in effect absent. The plate current, as shown on the meter, assumes a steady value. If the grid switch is now moved up, to put a negative bias on the grid, the plate current drops immediately. This is because the negative grid repels some of the negative electrons streaming through it. As the bias is increased, the plate current keeps falling until it reaches zero, at which point further increase of the negative charge on the grid has no effect; all the electrons from the cathode simply are being held back in the form of a cloud between the grid and the cathode. This point is called the *plate-current cutoff*, or just *cutoff*.

If making the grid negative retards the electron flow, it stands to reason that making it positive will increase it. This is easily shown by moving the switch to the positive leg of the bias supply. The plate current jumps, because the positive grid helps the positive plate to attract negative electrons from the cathode. Beyond a certain value of positive bias no further rise in plate current occurs, because the powerful combination of the two positive charges is sucking all available electrons out of the cathode. This condition is called *saturation*.

The grid's valvelike control of the plate current means that a much smaller change in grid bias is needed for any particular change in plate current than a change in plate voltage that would give the same effect. This is due merely to the fact that the grid is closer to the cathode than the plate is.

In practical circuits positive bias is rarely if ever used because it has the unwanted side effect of making the grid-cathode side of the triode act as a rather low value of resistance. In most cases negative bias—to repel more or fewer electrons—is used instead, as shown in B of Fig. 3-5.

If, in addition to this bias, a varying AC signal voltage from any source is applied to the grid, the plate current varies with the signal voltage. Because a small signal voltage applied to the grid can control a relatively large flow of plate current, the voltage variations appearing in the plate circuit are a magnified

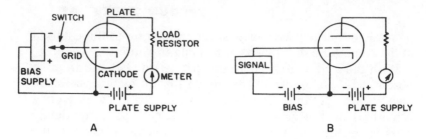

3-5. Making the grid of a triode negative or positive, as in (A), decreases or increases the plate current. In (B), the addition of signal voltage changes the control effect of fixed negative bias, and the signal is reproduced in amplified form in the plate circuit.

version of the signal applied to the grid. In other words, a triode has the ability to amplify a weak signal voltage.

Because most signals are very weak, a single amplifier tube or *stage* is rarely sufficient. Practical amplifiers for both audio and radio frequencies usually consist of several stages of virtually identical construction in an arrangement called *cascading*. The amplified output of the first tube feeds into the grid (input) circuit of the second, and the amplified output of the latter goes into the input of the third stage. It is not uncommon to find as many as eight or ten stages of amplification in amateur receivers and transmitters of various types.

Multigrid Tubes

Radio engineers learned quickly to appreciate the usefulness of the grid in the basic triode and they soon produced numerous multigrid types for particular effects or purposes. The most widely used ones are the *tetrode*, with two grids, and the *pentode*, with three.

The Tetrode   Since the cathode, the grid, and the plate of a triode each acts as the plate of a small capacitor, measurable capacitance exists between grid and plate, grid and cathode,

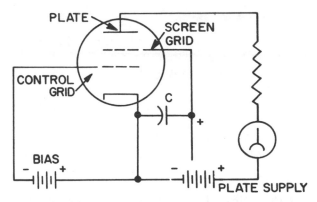

3-6. The screen grid in a tetrode acts as an electrostatic shield between the control grid and the plate.

and cathode and plate. These capacitances are known as *interelectrode capacitances.* The capacitance between the grid and the plate is usually the most critical since it produces undesirable coupling between the input circuit (the circuit between grid and cathode) and the output circuit (the circuit between plate and cathode). If the grid-plate capacitance becomes too large, high-frequency input signals may be short-circuited between the grid and the plate.

To reduce the grid-to-plate capacitance, a second or *screen grid* similar in construction to the control grid is inserted between grid and plate. A tube with a screen grid is called a *tetrode.* As shown in Fig. 3-6, a positive potential slightly lower than that on the plate is applied to the screen grid. This accelerates the electrons emitted from the cathode. Some of these electrons strike the screen grid, producing a slight current which is returned to plate supply. As a rule, screen current serves no useful purpose. Most of the electrons from the cathode pass through the open mesh of the screen grid to be collected by the more positive plate. The screen grid serves mainly as an electrostatic shield between grid and plate to reduce the capacitance between them. The screen acts more effectively when a bypass capacitor C is connected to the cathode. The gridplate capacitance of a typical tetrode is very small (less than 0.05 picofarad [pf] compared to that of a typical triode (usually greater than 2 pf).

The screen grid also makes the tetrode a better amplifier than a triode. Because of its presence in a tube, plate-voltage variations have little effect on the flow of plate current and the control grid has almost complete control over the flow of plate current. A typical tetrode can amplify its input signal as much as 800 times, whereas a triode may achieve·an amplification of only 30 times.

The Pentode   When high-velocity electrons strike the plate, they are likely to dislodge other electrons from it. This is called *secondary emission*. When this happens in a triode, the negatively charged control grid repels the displaced electrons back to the positively charged plate. In a tetrode, however, the positive screen grid attracts the displaced electrons and a reverse current flows between the plate and the screen grid. This reverse current reduces the total plate current. The effect of secondary emission is overcome by placing a third grid of the same construction as the control and screen grids, called a *suppressor* grid, between the screen grid and the plate. When this is done, the tube is known as a *pentode*. As shown in Fig. 3-7, the suppressor grid is generally connected to the

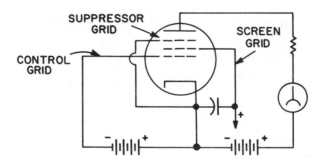

3-7. The suppressor grid of a pentode, between the plate and the screen grid, reduces secondary electron emission from the plate.

cathode, either internally or externally. It increases amplification and combats secondary emission by repelling electrons
back to the plate before they reach the screen grid.

Multiunit Tubes

To reduce the number of individual tubes in electronic equipment, the electrodes for two or more tubes are placed within
one evacuated envelope. Sometimes separate cathodes are
used for each tube section, but often a common cathode is
shared by the tube sections. In placing more than one tube in

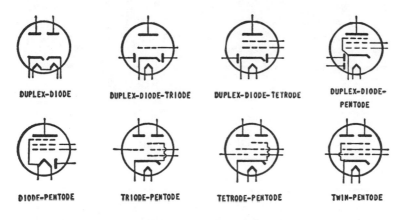

DUPLEX-DIODE   DUPLEX-DIODE-TRIODE   DUPLEX-DIODE-TETRODE   DUPLEX-DIODE-
                                                                               PENTODE

DIODE-PENTODE   TRIODE-PENTODE   TETRODE-PENTODE   TWIN-PENTODE

3-8. Schematic diagrams of some multiunit tubes.

a single envelope, some disadvantages result, such as increased interelectrode capacitance. However, when the disadvantages are not critical to the particular circuits involved, the
use of multiunit tubes results in more compact equipment.
Multiunit tubes are labeled according to the tube types they
contain. Figure 3-8 shows schematic diagrams of a number of
multiunit tubes.

Handling Tubes

It is important to know that all radio tubes fit in their sockets with a plugging action. To remove one, you must pull straight up; to replace it, you must line up the pins with the holes in the socket and then press straight down. Never, *but never*, attempt to loosen a tube by twisting it as you would an ordinary electric bulb. This is a sure way of damaging or even breaking the pins.

3-9. To ease a tight tube out of its socket, insert a thin screwdriver blade under its base and twist the handle just a little.

3-10. Then grasp the top of the tube firmly with dry fingers, pull without turning, and it will pop up without damage.

Tube Types

After more than half a century of development and refinement, vacuum tubes exist in about 3,000 different types, shapes, and sizes. Some representative tubes that a radio amateur is likely to encounter in experimental work are shown in Fig. 3-11.

Tubes are internationally standardized. Those of the same type number made in the United States, the Orient, and most of the countries of Europe are completely interchangeable.

3-11. Representative tubes found in amateur equipment: (A) and (B) glass envelopes, with plate connections to top caps, (C) standard glass type, (D) two sizes of metal envelopes, (E) metal type, with grid connection to top cap, and (F) and (G) miniature all-glass types.

4

Solid-State Devices

By definition, a *conductor* is a low-resistance material that allows electricity to flow readily, and an *insulator* is a high-resistance material that blocks current altogether. Conductors do offer some resistance, depending on their exact composition and their physical dimensions, but this is generally an incidental characteristic. The actual difference in resistance between common conductors and insulators is enormous. For example, silver has a resistance of only one-millionth of an ohm between any two faces of a cube measuring a centimeter on a side, while a similar block of mica has a resistance of about a million million ohms, a value so high that it represents in effect an open circuit.

There is a sort of twilight zone between conductors and insulators, and the materials that fall in it, mostly natural or man-made crystals, are known as *semiconductors*. For example, a centimeter cube of pure germanium measures only about 50 or 60 ohms; a cube of pure silicon, 50,000 to 60,000 ohms. These resistances fall to much lower values if certain chemical "impurities" are added to the crystals. What is significant about semiconductors treated in this manner is that some of their internal electrons apparently float around loosely, and can be made to move under the influence of very low applied voltages. While conventional vacuum tubes for receiving purposes require plate voltages from about 75 to 300 volts, typical semiconductors in similar applications need only between 1½ and 12 volts. Since the controllable electron stream in a tube flows through a vacuum, while in semiconductors it goes through a solid, the basic term *solid-state* has been adopted to distinguish semiconductors from tubes.

The major part of the electric energy supplied to most tubes is consumed by the heater element that boils electrons out of a cathode. (See Chapter 3, dealing with vacuum tubes.) Solid-state devices do not need thermal priming; their loose electrons are on tap at all times and go to work the instant an external voltage is applied. Semiconductors are therefore smaller than tubes, require less space, wiring, and operating power, and work in simpler circuits.

Although solid-state technology is generally considered a development of the 1950's, it actually dates back to the turn of the century. As early as 1903, an American experimenter named Greenleaf Whittier Pickard investigated the possibilities of certain crystals as detectors (that is, rectifiers) of radio signals. In 1906 he obtained excellent results from silicon, which today is a favored material for many solid-state devices. Numerous other crystals, including even ordinary coal, were used successfully. The most sensitive was found to be *galena* (chemically, lead sulphide), a cheap and abundant by-product of silver-mining operations in the western part of the United States.

The Transistor

Until 1948 all solid-state devices were essentially one-way conductors, and could be used only as signal detectors and as rectifiers in AC power circuits. Unlike tubes, they could not amplify weak signals or act as oscillators to produce high-frequency alternating current for transmission and other purposes. However, in 1948 the electronic art was literally set on its ear by the introduction of an entirely new semiconductor device called the *transistor*, which could do everything the tube could do, and more, within certain power limitations. A product of intense, highly organized team engineering in the vast Bell Telephone Laboratories, the transistor was an overnight sensation. Requiring no heater or filament current and no glass bulb or vacuum, and taking the form of strong metal beads the size of match heads or peas, the transistor obviously was ideally suited for a wide variety of electronic equipment ranging from tiny hearing aids to portable receivers and transmitters and computers and space instruments.

Semiconductor Theory

There is no single, universally accepted answer to the question "How do semiconductors, and particularly transistors, work?" Several theories have been advanced, and they differ not only in basic approach but also in mere terminology. This is not surprising in view of the fact that the very nature of electricity is still a matter of widespread speculation among scientists. Although the operation of the vacuum tube clearly supports the concept that electricity is a movement of negative electrons toward the positive side of a circuit, some transistor texts confuse the student by indicating both "electron flow" and "conventional current flow"—in the other direction—in the same hookup!

Using the electron theory, it is possible to offer a reasonable explanation of semiconductor action. Readers with a college background in modern physics are referred to the advanced books listed in the Appendix.

The atom may be pictured as a central core or nucleus having a positive electric charge, surrounded by a cloud of orbiting electrons having a negative charge. The electrons nearest the core are held more firmly by the latter's positive charge than those farther out at the edges, but under normal circumstances the positive and negative charges balance and no electrons escape. The outer electrons, more easily torn loose, are called *valence* electrons.

The atoms of pure semiconductors are arranged in a crystalline structure, an orderly framework called a *lattice*. In the lattice, atoms line up so that their valence electrons are shared. Valence electrons of adjacent atoms are bound together to form *electron-pair bonds*. This is shown in simplified fashion in Fig. 4-1. These bonds are quite tight, and there are no free electrons on tap to be influenced by outside electric charges. Thus, in effect, the lattice has high resistance.

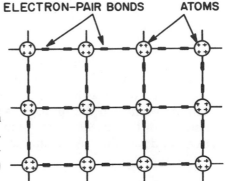

ELECTRON–PAIR BONDS   **ATOMS**

4-1. Crystal lattice structure of a pure semiconductor material. (*Courtesy of the RCA Semiconductor and Materials Division.*)

It is possible to split the electron-pair bonds and to free some electrons by applying heat or high voltage, but these measures are awkward and troublesome. The big breakthrough in semiconductor technology came with the discovery that the same effect could be accomplished much more simply by adding "impurities"—extremely small amounts of other elements having different atomic structure—to the pure lattices. This process is called *doping*, and is the most critical part of semiconductor manufacturing because the ratio of impurities to pure materials is something like one part in ten million!

Doping works in two directions. If the added impurity element has more valence electrons than the pure semiconductor material, the extra electrons tend to float loosely within the lattice because there are no unpaired electrons available in the lattice with which they can form new electron-bond pairs. See Fig. 4-2. Loose electrons are easily affected by an outside charge, so electrons can readily be made to flow; in effect, the loose electrons give the doped semiconductor material a resistance lower than that of the previous pure form. A material having this excess-negative characteristic is called n-type. Common n additives are arsenic and antimony.

Impurity atoms such as aluminum, gallium, and indium have *fewer* valence electrons than semiconductor atoms. There are not enough of them to form complete electron-bond pairs with all of the latter's valence electrons, so they leave what amounts to holes (that is, areas without paired electrons) in the lattice structure. See Fig. 4-3. Some electrons of adjacent pair bonds tend to shift from their positions under the influence of outside charges and to move into the holes. This movement of electrons constitutes a flow of electricity. The initial vacancy in the lattice is said to have a positive charge because of the absence there of negative electrons, so semiconductors doped in this manner are called p-type.

p-n Junctions

The basic solid-state device consists of a combination of p-type and n-type materials in simple contact, as in Fig. 4-4; this is called a p-n *junction*. At the junction itself, some of the loose electrons in the n-type tend to diffuse into the adjacent holes. The holes thus acquire a slight negative charge, while the previously all-negative area at the junction becomes slightly positive because it has lost some of its electrons to the holes. This intermediate area is called the *space-charge region, transition region,* or *depletion layer.* The very slight electric charge here can be represented as an imaginary battery, as shown in Fig. 4-5. It is known as the *energy barrier* because it discourages further diffusion across the junction; that is, the initial negative charge acquired by the holes in the space-charge region prevents additional electrons from the n-type material from crossing into more holes in the p-type.

The condition just described continues to exist only as long as the p-n junction is isolated. When external voltages are applied, the nature of the space-charge region changes mark-

4-2. Lattice structure of n-type "doped" semiconductor. (*Courtesy of the RCA Semiconductor and Materials Division.*)

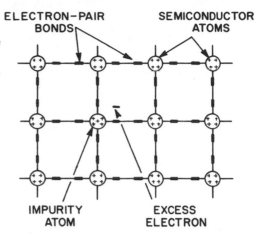

ELECTRON–PAIR BONDS   SEMICONDUCTOR ATOMS

IMPURITY ATOM   EXCESS ELECTRON

4-3. Lattice structure of a p-type "doped" semiconductor. (*Courtesy of the RCA Semiconductor and Materials Division.*)

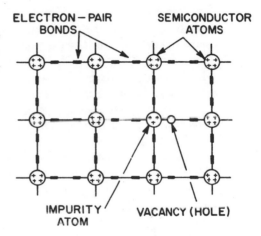

ELECTRON–PAIR BONDS   SEMICONDUCTOR ATOMS

IMPURITY ATOM   VACANCY (HOLE)

4-4. Interaction of electrons and "holes" in the space-charge region of a p-n junction. (*Courtesy of the RCA Semiconductor and Materials Division.*)

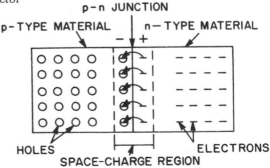

p-n JUNCTION

p-TYPE MATERIAL   n-TYPE MATERIAL

HOLES   ELECTRONS

SPACE-CHARGE REGION

edly. Consider first the simple circuit of Fig. 4-6, which shows a battery connected to the p and n ends of a junction. The free electrons in the n-type, being negative, are drawn away from the material toward the positive side of the battery. This loss of electrons tends to make the material more positive than before, in effect widening the positive side of the space-charge region. Simultaneously, electrons from the negative pole of the battery go into the positive p-type material, diffuse through the holes, make this section more negative than before, and in effect widen the negative side of the space-charge region. The total effect is to make the latter so wide that it is no longer the imaginary battery shown in Fig. 4-5 but assumes the characteristics of a real battery having a voltage almost equal to that of the external battery. A condition of voltage balance sets in, and, as a result, there is virtually no current flow through the circuit, as indicated by the thin arrow in Fig. 4-6. A p-n junction with the battery polarity as shown in this diagram is said to be *reverse-biased*.

If the external battery is switched around, as in Fig. 4-7, electrons in the p-type break out of their electron-pair bonds under the pull of the positive side of the battery, creating new holes in the material, and they travel toward the battery. This loss of electrons makes the p-type material more positive than before and causes it to attract more electrons through the space-charge region from the n-type. As these electrons move across, they are replaced by other electrons from the negative side of the battery. In effect the space-charge region virtually disappears, the energy barrier is no longer a barrier, and electrons flow merrily around the circuit, as indicated by the heavy arrow in Fig. 4-7. A junction with this battery polarity is said to be *forward-biased*.

If we substitute a source of alternating current for the batteries of Figs. 4-6 and 4-7, it is easy to see that a p-n junction is a simple rectifier. While there is some current flow in the reverse-bias condition, this is so small in comparison with the heavy current in the forward-bias mode that it can be disregarded in practical applications.

4-5. Voltage effect at the center of a p-n junction. (*Courtesy of the RCA Semiconductor and Materials Division.*)

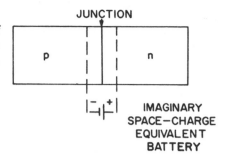

IMAGINARY SPACE–CHARGE EQUIVALENT BATTERY

4-6. When biased in this manner, a p-n junction has a high resistance and permits only a very small electron flow. (*Courtesy of the RCA Semiconductor and Materials Division.*)

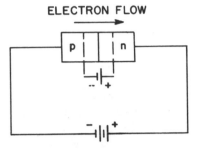

4-7. With the bias of this polarity, the p-n junction offers low resistance and passes a heavy current. (*Courtesy of the RCA Semiconductor and Materials Division.*)

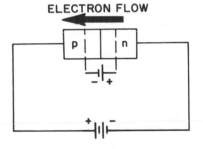

Transistor Operation

Another way of describing p-n action is to say that a junction has very high resistance with reverse-bias and very low resistance with forward-bias. Recall from Ohm's law that the power developed by any current is greater in high resistance than in low resistance; that is, power in watts equals current in amperes squared times resistance in ohms. An *increase* in power from one circuit to another (*amplification* or *gain*, as obtained with vacuum tubes) is therefore possible if the input or control circuit has low resistance and the output circuit has relatively higher resistance, and *if the current transferred from one to the other is maintained with little or no loss.* The solid-state device that accomplishes this difficult trick is the remarkable *transistor.*

A transistor in basic form has three semiconductor elements, as shown in Fig. 4-8. For purposes of discussion let's make the ends of n-type. These form a sandwich with a very thin filling of p-type material, so the device is called an n-p-n structure. By means of suitable batteries, the left-hand input n-p junction is forward-biased and therefore has low resistance, while the p-n output junction is reverse-biased and its circuit has high resistance. In the input section, electrons flow readily from the n wafer to the center p wafer, because the latter is made additionally positive by the biasing battery. However, instead of returning to the latter, they mostly diffuse through the p section to the output n wafer, where they are attracted by the positive charge from the right-hand battery. In actual transistors as much as 99½ percent of the electron current exists through the right-hand n semiconductor; the small remainder completes its circuit in the left-hand n section. This current passing through the high resistance of the output circuit represents a very much higher power than the same current if it flowed through the low resistance of the input circuit. In practical terms this means that a properly biased transistor can offer enormous amplification; that is, a weak signal impressed on the left-hand n-p junction reappears as a strong signal in the right-hand p-n junction.

If the transistor sandwich is made of one negative and two positive semiconductors, it is called a p-n-p structure. With the battery polarities reversed to match this arrangement, the operation of the device is similar to that of the n-p-n. Both of these types are in widespread use. Figure 4-9 shows schematic symbols for the two types.

To distinguish transistor elements from each other, the three regions are designated as the *emitter, base,* and *collector.* In normal practice the emitter-to-base junction is forward-biased, and the collector-to-base is reverse-biased. In certain

4-8. Basic transistor construction and operation. The heavy arrow indicates a strong diffusion of electrons from the input to the output circuit; the thin arrow indicates a very small flow in the input section alone. (*Courtesy of the RCA Semiconductor and Materials Division.*)

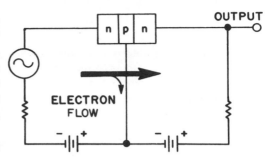

4-9. Schematic symbols for two standard types of transistors: (A) n-p-n, (B) p-n-p. (*Courtesy of the RCA Semiconductor and Materials Division.*)

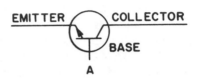

special applications where transistors are used as coupling devices and actual amplification is not needed, the biases are adjusted for minimum diffusion effect.

There are literally thousands of types, sizes, and shapes of transistors, and the number grows daily as scientists produce esoteric new semiconductor materials.

Semiconductor Advantages and Disadvantages

It would seem that solid-state devices are the answer to most of the problems of communication equipment design. This is not quite true. One of the unfortunate claims made for transistors is that they last forever, or almost forever, because they have no filaments to burn out or vacuums to lose. The truth of

4-10. Three typical transistors and a tiny tunnel diode. Some transistors are plugged into small sockets, in the same manner as tubes, but more often they are simply soldered into holes in printed circuit boards.

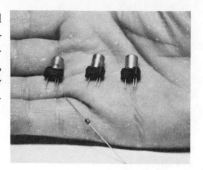

the matter is that they can be damaged quite readily by too much heat, applied, for example, when connecting leads are soldered. This too is not surprising, since heat by definition is the energy of the motion of atoms, and excessive motion can destroy the delicate crystalline structure of solid-state lattices.

Excessive voltage or voltage of the wrong polarity is a common cause of trouble in much equipment. Like heat, this tends to upset the lattices. In a dry room, a person walking on a wool or nylon rug can generate enough static electricty to disable a transistor radio or amplifier when he touches it.

Countering the disadvantages are undisputed advantages for many applications. With no filaments and no warm-up time, transistor devices snap into action without delay. For all except power-amplifier stages, they work on such low voltages and currents that tiny batteries can energize them satisfactorily for long periods; furthermore, the batteries can be built right into the equipment.

For fixed-station applications, where space is no object and unlimited energy is available from an AC outlet, transistors offer no particular advantages over tubes. Some of the best known amateur gear on the market is still all-tube, with the exception of a few small diodes as signal rectifiers. A hybrid arrangement now in widespread use, particularly for transmitters, consists of solid-state devices in the low-energy circuits and a pair of husky tubes of proven dependability in the power-amplifier (output) positions.

Portable equipment, of course, is another story. About the

only limitation on their size is the size of the batteries needed to power them. Many hand-held transceivers for 144 megahertz and higher frequencies are actually smaller than many cameras of the "instant" type.

4-11. These are only some of the solid-state devices used in a typical ham transceiver. Along the top are relatively large transistors of medium-power rating. At the bottom is what looks like a colony of insects; actually, these are small-signal transistors with wire leads. In the center is a collection of integrated circuits ("IC's"). These are combinations of highly miniaturized transistors, capacitors, resistors, etc., all connected in manufacture to function in a variety of applications in receivers, transmitters, and other electronic assemblies.

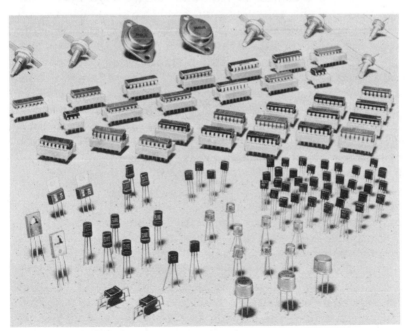

5

Receiver Theory

Receiver Tuning

In Chapter 2 we discussed resonant circuits consisting of inductance (L) and capacitance (C). We learned that these circuits are selective to certain AC frequencies, depending on the values of L and C used. That is, any series or parallel circuit of inductance and capacitance has a resonant frequency at which the values of the inductive and capacitive reactances (X_L and X_C) are equal. Series and parallel circuits of L and C have interesting characteristics at resonance. A series L-C circuit with the voltage of its resonant frequency impressed across it offers minimum impedance to the flow of current. On the other hand, a parallel L-C circuit with the voltage of its resonant frequency impressed across it offers maximum impedance. The point of resonance may be sharply or broadly defined, depending on the resistance of the inductive component used or on the value of external resistors introduced into the circuit for the deliberate purpose of broadening the tuning.

The phenomenon of series and parallel resonance plays a very important part in radio since, among other things, it is responsible for receiver tuning. Here's how it works.

We have, as diagrammed in Fig. 5-1, a parallel L-C circuit connected between ground and an antenna. The capacitor of the parallel L-C circuit is variable. (Resonant frequency of an L-C circuit may be changed by varying either its inductance or its capacitance.) Imagine, then, a radio wave of a certain frequency striking the antenna and inducing in it an AC voltage which causes a minute AC current to flow through the parallel L-C circuit to ground. If the frequency of the radio wave is the same as the resonant frequency of L and C, the parallel L-C circuit in Fig. 5-1 offers maximum impedance to the flow of the AC current and a small AC voltage appears across it. When radio waves of other frequencies strike the antenna, the induced AC currents produce no appreciable voltage drop across the L-C circuit since it offers practically no impedance to AC currents not at its resonant frequency. In other words, an AC voltage representative of the radio waves striking the antenna appears across the L-C circuit only when the radio waves have the same frequency as the resonant frequency of the L-C circuit. In this manner the parallel L-C circuit can select radio waves of one frequency over radio waves of other frequencies that strike the antenna. Since the capacitor is variable over a range of values, we can change the resonant frequency of the L-C circuit and thereby select or "tune in" radio waves over a range of frequencies.

Another aspect of the operation of parallel L-C circuits may be seen by referring to Fig. 2-21, which shows the effect of resistance on the sharpness of resonance. Minimum resistance

5-1. A parallel resonant L-C circuit used to select or "tune in" a radio signal.

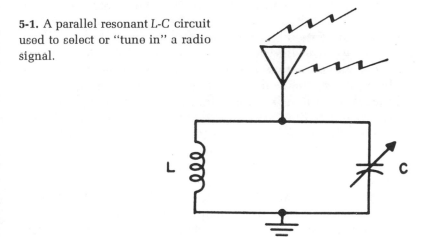

provides maximum sharpness or most selective tuning of the desired frequency to the exclusion of higher or lower frequencies.

The frequency that the L-C circuit tunes in or selects depends on its resonant frequency, which in turn depends on the values of L and C. Resonant frequency (f_r) may be determined as

$$f_r = \frac{1}{2\pi\sqrt{LC}}$$

where f_r is the frequency in cycles per second, L the inductance in henries, and C the capacitance in farads.

The above formula also tells us that as L and C become larger, the resonant frequency becomes lower. Conversely, as L and C become smaller, the resonant frequency becomes higher. For example, a parallel L-C circuit for tuning the relatively low frequencies of the broadcast band (550 to 1,600 kHz) might have a coil (L) with 100 or so turns of wire, and a variable tuning capacitor (C) with 20 intermeshing plates. On the other hand, an L-C circuit for tuning the 14,000- to 14,350-kHz amateur band might have a coil with only 10 turns and capacitor with 6 intermeshing plates.

Figure 5-2 shows an antenna, parallel L-C circuit, and first stage of RF amplification that might be used for a receiver. Notice that the antenna is not directly connected, but is inductively coupled to the L-C circuit. This is usually done to isolate the antenna from the first RF amplifier stage. When a radio signal of the same frequency as the resonant frequency of the L-C circuit strikes the antenna, an AC voltage representative of the radio signal appears at the control grid of V-1. The amplified version of this signal then appears in the plate circuit of V-1 across the RF transformer T_1. The secondary of T_1 then applies the signal to the grid of the next RF stage for further amplification. When the signal has been sufficiently amplified, it is applied to a detector. This arrangement is called *tuned radio-frequency* amplification (TRF).

Since vacuum tubes and transistors perform virtually identical circuit functions, the theory presented in this chapter applies to solid-state receivers as well as to conventional tube sets.

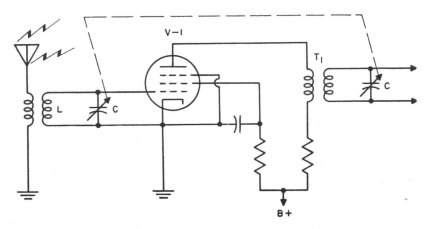

5-2. Tuning and the first RF amplifier stage of a receiver.

Detection

Most amateurs use a method of voice modulation known as *amplitude modulation* (AM). An important variation of this system, called *single-sideband suppressed carrier*, is treated in detail in chapter 6.

The action of amplitude modulation on a radio-frequency carrier wave is shown in Fig. 5-3. Notice that the modulator imposes the voice intelligence on both the positive and the negative portions of the carrier wave, as indicated in Fig. 5-3(B). In detecting an AM signal, two things are accomplished: (1) the modulated carrier is rectified so that only the positive portion remains, and (2) the RF carrier is removed so that only the intelligence as represented by the carrier-wave peaks remains, as indicated in Fig. 5-3(C). A number of circuits have been designed for AM detection.

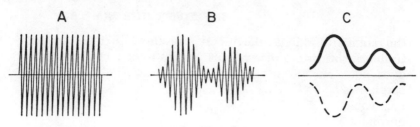

5-3. The action of amplitude modulation on an RF carrier: (A) an unmodulated carrier, (B) a modulated carrier, and (C) the detected audio intelligence.

The Diode Detector

The plate of diode detector (Fig. 5-4) receives a modulated RF signal from the RF amplifier section of a receiver. The action of a diode detector is essentially that of a half-wave rectifier. In other words, when the modulated signal at the plate swings positive, the diode conducts and a rectified current flows through the load resistor R. Since the output of a rectifier is directly proportional to the input, the current flowing through R causes the voltage drop across R and the capacitor C to vary in accordance with the amplitude of the modulated input signal. The value of C is such that the voltage variations across R and C do not follow the rapidly changing RF signal, but only

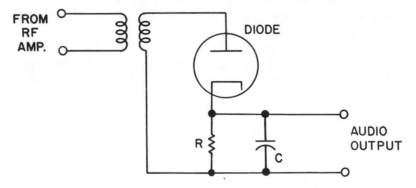

5-4. A simple diode detector.

the rectified peaks, that is, the audio intelligence. The value of C should be such that it filters out the radio frequency but does not alter the audio variations. A set of very sensitive earphones could be connected to the audio output and the signal would be heard. However, the output of the detector is normally fed to one or two stages of audio amplification to strengthen it for more convenient listening.

The Grid-Leak Detector

The grid-leak detector (see Fig. 5-5) uses a triode. The modulated RF signal is first impressed across the grid and cathode. This causes the grid to act like the plate of the diode detector described above. The grid-leak resistor R serves as a load and C_1 as an RF bypass for this grid-cathode rectifying section. As the RF signal on the grid swings positive, current flows from the grid to the cathode through R. The voltage drop across R makes the grid negative with respect to cathode; i.e., the rectified signal between grid and cathode serves to bias the triode. The amplitude of this bias varies directly with the peaks of the incoming RF signal. Since the triode's plate current varies with its grid bias, the voltage appearing in the output of the triode is an amplified and rectified version of the modulated RF signal. Thus, the grid-leak detector serves two purposes. The grid-cathode portion of the circuit forms a sim-

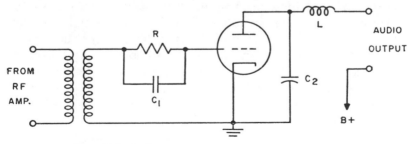

5-5. A simple grid-leak detector.

ple diode rectifier, while the cathode-grid-plate combination acts as a vacuum-tube amplifier.

The RF choke L and the capacitor C_2 serve to filter out the RF and leave only the audio signal at the output. The high impedance of L at high frequencies blocks the path of RF from the output while the low impedance of C_2 at high frequencies shorts the RF past the output to ground; L and C_2 have exactly the opposite effect on the low-frequency audio signal. Here L has relatively little effect and C_2 offers such a high impedance that the audio signal is not shorted out. This combination of L and C_2 is a standard filter arrangement for blocking a high-frequency RF signal and passing a low-frequency audio signal. Earphones may be connected to the output of this grid-leak detector or audio amplification may be added to drive a loudspeaker.

The Regenerative Detector

A simple regenerative circuit (Fig. 5-6) usually consists of a grid-leak detector in which a portion of the output signal is fed back to the input circuit of the tube. When a modulated radio-frequency signal is applied to this circuit, the tube acts as a grid-leak detector and a detected and amplified signal appears in the plate circuit. However, by means of L_1, a coil inductively coupled to the secondary winding L of the RF input transformer, a portion of the detected signal in the plate circuit is fed back to the grid circuit. This feedback reinforces the original signal, which now makes the plate signal stronger than before. The amplified plate signal feeds still stronger impulses to the grid, and the process keeps going—not, however, without limit. If the coupling between L_1 and L is too close, overly strong impulses from the plate circuit shock the circuit consisting of L and tuning capacitor C into oscillation on its own accord. The RF energy thus generated mixes with the incoming AM signals, and the net result is whistling and distortion.

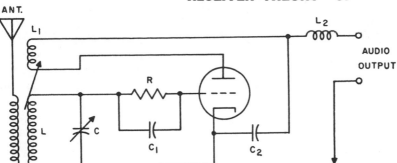

5-6. A simple regenerative detector.

The coil L_1 is called a *tickler*. The arrow drawn between it and L in Fig. 5-6 indicates that the physical coupling between the two is variable, to control the amount of feedback.

Feedback can be accomplished capacitively as well as inductively.

In Fig. 5-6 L_2 and C_2 remove the RF component of the detected signal in the same manner as L and C_2 in the straight grid-leak detector shown in Fig. 5-5. The audio output of this regenerative detector may be fed to sensitive earphones or to a stage or two of audio amplification.

Simple regenerative receivers using the exact circuit of Fig. 5-6 were used by hams for many years before the advent of the far superior superheterodyne type. They are now completely outmoded. However, the principle of regeneration still has important applications in communication equipment.

The Superregenerative Detector

The regenerative detector is limited in the amount of amplification it can produce. The *superregenerative* detector stretches this limit by introducing into the circuit an AC (usually between 20 and 200 kHz) which *quenches* the oscillations. This quenching voltage interrupts the operation of the regenerative detector by driving the detector to cutoff every time the quench goes negative. Thus, by effectively turning the detector tube on and off at a very rapid rate, it is prevented from oscillating. The detector itself, or a separate tube, may be used to furnish the quenching voltage. In either case, the quench must operate at a frequency above audibility; otherwise the noise generated would drown out the incoming modulated signal.

Superregenerative receivers tend to be noisy and broad in tuning, and never achieved the popularity of the straight regenerative type. Like the latter, they are now obsolete, but the usefulness of the principle continues.

The Superheterodyne

Straight TRF receivers tune very broadly and amplify unevenly, and regenerative sets are critical and unstable. Both types are obsolete, but certain of their more desirable features are often found in the *superheterodyne* ("superhet" for short), a circuit universally used in receivers for communications, television, high fidelity, and AM and FM purposes.

In the TRF receiver, the carrier frequency of the incoming signal is never changed. In the superheterodyne, all incoming carrier frequencies are converted to a lower frequency called the *intermediate frequency*. This intermediate frequency (IF) is always the same regardless of the frequency of the incoming signal. By changing the various incoming frequencies to a single IF, the subsequent stages of amplification before detection can be made to operate very efficiently. The nonlinear amplification characteristic and the cumbersome ganged tuning of three or four RF amplifier stages featured in the TRF receiver are eliminated. Furthermore, the superheterodyne receiver with its principle of IF amplification provides stable operation with high sensitivity and selectivity. Here's how a superheterodyne works.

Let us assume that we have two tuning forks, one with a

frequency of 100 hertz and the other with a frequency of 400 hertz. If we strike these forks simultaneously, we hear the following: (1) a 100-hertz tone from one fork, (2) a 400-hertz tone from the other, (3) a 500-hertz tone which is the *sum* of the two frequencies, and (4) a 300-hertz tone which is the *difference* between the two frequencies. In other words, when two forks of different frequencies are struck simultaneously, four different frequencies result—the two original frequencies, and their sum and their difference. This mixing of two frequencies is known as *heterodyning* or *beating*, and the new frequencies are *heterodynes* or *beats*.

If we apply this principle of heterodyning to two AC signals of different frequencies we get the same results. For example, suppose we have a tuned circuit which is receiving a 1,000-kHz carrier. In addition we also have a radio-frequency signal oscillator which generates a 1,455-kHz signal. If the two signals are electronically mixed, two new frequencies result: a 2,455-kHz signal, and a 455-kHz signal.

To continue a little further, let us also assume that the variable capacitor in the tuning circuit is mechanically coupled to the variable capacitor which determines the output frequency of the local oscillator, so that when the two signals are mixed the *difference frequency always remains the same*. In the previous example, when the tuning circuit was receiving 1,000 kHz, the signal generator produced a 1,455-kHz signal and the difference frequency was 455 kHz. If we now change the tuning capacitor to receive a 1,370-kHz carrier, the variable capacitor of the signal generator (being mechanically coupled or ganged to the tuning capacitor) likewise changes to make the generator's output 1,825 kHz. As a result, the difference frequency (1,825 kHz minus 1,370 kHz) would still be 455 kHz. In other words, the difference frequency is always the same (in this case 455 kHz) for any carrier frequency tuned in by the tuning capacitor.

Figure 5-7 is a block diagram of a basic superheterodyne receiver. Converting the incoming signal to the intermediate frequency is accomplished in the *first detector* or *mixer* stage.

5-7. Block diagram of a basic superheterodyne receiver (power supply not shown).

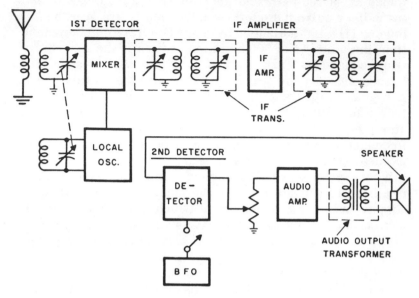

The incoming carrier selected by the antenna tuning circuit is applied to one of the grids of the multigrid mixer tube, while the output of the local oscillator is applied to another grid. The local oscillator and the incoming signal are thus electronically mixed. At the plate of the mixer tube four frequencies are present—the two originals and the sum and difference frequencies. The IF transformer coupling the output of the mixer to the first stage of IF amplification is tuned so that only the difference frequency is applied to the grid of the first IF amplifier tube. Two stages of IF amplification furnish enough gain to make the receiver fairly sensitive. Sometimes a small amount of regeneration is introduced to boost the amplification. However, for increased sensitivity, more IF stages can be used. Superheterodyne receivers can also be made more sensitive and selective by the addition of a stage of tuned RF amplification ahead of the mixer.

In some receivers the mixer and local oscillator functions are performed by two separate tubes; in other sets the functions are combined in a multielement tube called a *pentagrid converter*.

Second Detector and AF Amplifier

The output of the last IF stage is at a much lower frequency than the original AM signal picked up by the antenna, but it is still modulated RF and it must still be detected. This is done in the second detector stage, which is followed by a conventional AF amplifier for eventual reproduction of the signal by a loudspeaker or headset.

Notice in Fig. 5-7 the use of an audio transformer to couple the output of the audio amplifier to the speaker. This transformer provides for efficient transfer of energy from the relatively high-impedance plate circuit to the low impedance of the speaker's voice coil. Since the voice coils of most speakers range from about 4 to 16 ohms, this matching is important. With high-impedance earphones (2,000 ohms or more), audio transformers are not required, although a coupling capacitor must be used to isolate the earphones from the amplifier's high-voltage DC plate supply.

CW Reception

If CW signals are tuned in by a typical superhet as just described, the end result in the loudspeaker is a series of confusing clicks or thumps. A dot comes out as two clicks close together; a dash, as two clicks somewhat more separated. To obtain the pleasant whistling signals that make CW easy to follow, it is necessary to add another local oscillator and to couple it to the second detector. See Fig. 5-7. This is called a *beat-frequency oscillator* (BFO). It is tuned so that the frequency difference between it and the IF signal falls in the audio range, usually between about 500 and 1,000 hertz.

While the local oscillator used for conversion of carrier frequencies to the intermediate frequency, in the mixer stage, is also a BFO by strict definition, the term BFO is reserved in general practice for the CW signal converter.

Automatic Volume Control

Automatic volume control (AVC) is incorporated in all superhet receivers as an operating convenience for voice reception. It tends to keep the sound level at the speaker constant regardless of varying signal strength at the antenna. The usual AVC circuit operates by taking the average DC level of the detected signal and applying it as a control bias to the grids of the IF amplifier tubes. It works in the following manner. When the signal strength on the antenna increases, a corresponding increase appears at the second detector. However, as the audio level increases, the negative bias also increases because of the AVC action. Increasing the negative bias on the IF amplifiers reduces their output, to effectively reduce the signal at the detector. This action is almost instantaneous and, as a result, the volume of sound at the speaker tends to remain relatively constant. Amateur receivers are provided with a switch in the AVC circuit since it is a disadvantage for receiving very weak signals. Ultimate control of the volume of sound at the speaker is provided by a volume-control potentiometer which permits adjustment of the amount of the detected signal applied to the audio amplifier tube.

Double Conversion

Receivers of the type diagrammed in Fig. 5-7 are known as *single-conversion* superhets, because the carrier frequencies undergo only one change. However, there are many sets in which a second change takes place, and these are called *dou-*

ble conversion. There are even some triple-conversion jobs, but they are quite complicated and are mainly technical exercises for advanced experimenters.

As pointed out earlier, heterodyning in the first detector produces *two* beats, a difference frequency and a sum frequency. The IF amplifiers are resonated to the lower or difference frequency, but that does not mean that the sum frequency just disappears. It is still present, and it can produce spurious signals called "images" in the IF stages by beating on its own accord either with harmonics of the local oscillator or with random signals present with the desired signals in the relatively broad circuits ahead of the mixer stage. *Harmonics* are secondary frequencies in RF oscillators and are arithmetical multiples of the base or *fundamental* frequency to which the circuits are tuned. For example, if an oscillator is adjusted for a fundamental of 500 kHz, it is also quite likely to generate harmonics at 1,000, 1,500, 2,000, 2,500 kHz, and upward. Because there are thousands of shortwave stations on the air, some of this unwanted heterodyning is inevitable. It shows in a receiver as mysterious whistling signals, known appropriately as "birdies."

The greater the separation between the sum and difference frequencies, the less the chance of image interference. This is obtained easily by making the first IF higher than before, perhaps 1,500 to 4,000 kHz. Straight amplification on these frequencies is rather poor, so after a stage or two of this IF to filter out the birdies, the signal is converted back to a lower value, usually 455 kHz, which permits high amplification and selectivity.

Frequency Modulation Reception

Frequency modulation (FM) differs from AM in that its amplitude remains constant but its frequency changes. If an FM signal is fed into a conventional rectifying detector, as in Fig. 5-4 or 5-5, the output of the latter would be a clipped unidirectional current pulsating at the varying frequency of the signal, millions of hertz above the limit of human response.

For proper FM reception an entirely different kind of detector is needed. This must discriminate between deviations above and below the mean frequency of the carrier and translate these deviations into voltages whose *amplitude* varies at the audio frequencies of the voice signals. The circuits of FM detectors are very tricky.

Aside from their detectors, FM and AM superhets use similar circuitry.

5-8. Faced with the problem of choosing between a general-coverage receiver and a ham-band receiver, many hams buy both and have the best of two worlds. A typical station so outfitted is WA8ASQ, operated by Paul J. Kirsch, Livonia, Mich.

6

Transmitter Theory

In a radio transmitter, the circuit that generates the high-frequency AC current that ultimately produces radio waves from an antenna is called an *oscillator*. In most amateur transmitters the heart of the oscillator is either a tube or a transistor.* In addition to performing such a vital function in transmitters, oscillators are also found in a large variety of other electronic equipment.

Any tube or transistor amplifier can be made to oscillate or generate a self-sustained AC signal if more energy is fed back in phase from the plate circuit than is lost in the grid circuit. The frequency of the AC signal depends on the inductance and capacitance in the grid circuit.

The Grid-Leak Oscillator

The grid-leak oscillator shown in Fig. 6-1 is often referred to as a *tuned-grid* or *tickler-coil* oscillator. Here is how it works.

When the switch S is closed, the tube conducts. The grid, being in the path of the cathode-to-plate electron stream, collects a few electrons. Capacitor C_2 blocks the flow of DC and does not allow the excess grid electrons to flow back to the

---

* A basic triode tube is shown in this chapter for purposes of explanation because its elements are more easily identified in schematic diagrams than those of transistors. The circuit operation is the same regardless of what type of amplifier device is used. Think of the cathode of a tube as equivalent to the emitter of a transistor, the grid as the base, and the plate as the collector, and everything will be clear.

6-1. A grid-leak oscillator.

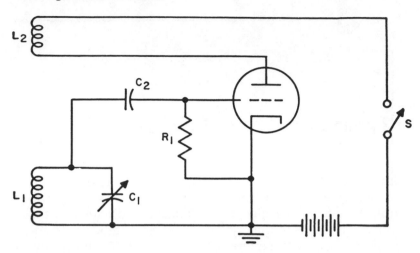

cathode. In order to return to the cathode these electrons must flow, or *leak*, through resistor R_1. This action causes a voltage drop across R_1 and places a negative bias on the grid. This grid action from positive to negative is enough of a triggering pulse to start the L_1/C_1 circuit oscillating. Once oscillations start, they are amplified by the tube and appear across L_2. This tickler coil L_2 is inductively coupled to the grid coil L_1. The feedback due to this coupling is sufficient to sustain oscillations. Capacitor C_1 is made variable so that the circuit can be tuned over a range of frequencies.

The Hartley Oscillator

The Hartley oscillator, shown in Fig. 6-2, is identical with the basic tickler circuit of Fig. 6-1, with the physical difference that it uses a single tapped coil instead of two separate coils. The "tickler" is now the lower section of L_1, between the tap and $B-$. It is directly in series with the plate-cathode elements of the tube. In effect L_1 is an autotransformer, with the lower part the primary and the entire winding the secondary. Capa-

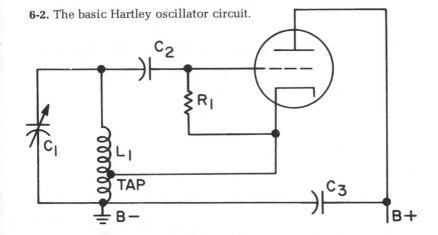

6-2. The basic Hartley oscillator circuit.

citor C_3 is merely a bypass to provide a low-impedance path for the amplified RF signal component of the plate current.

The circuit of Fig. 6-2 is probably the most widely used variable RF oscillator in amateur practice because it is extremely simple and reliable and because it permits the tuning capacitor C_1 to have its rotor plates grounded directly to the usual metal chassis of the actual transmitter or other piece of equipment.

The Colpitts Oscillator

This oscillator is similar to the Hartley, but uses capacitive instead of inductive coupling between the plate and grid circuits of the tube, as in Fig. 6-3. The effect of the tickler coil is produced by capacitor C_2, which is charged by part of the amplified RF signal in the plate circuit and then discharges through L_1 and C_1 in the grid. A disadvantage of this arrangement is the need for a double but split variable capacitor C_1/C_2. Variations of the Colpitts oscillator are useful for special purposes.

Since the feedback capacitor does not pass DC, the DC plate circuit from cathode to plate is completed through the RF choke L_2. The latter passes DC readily, but because it has a

6-3. The Colpitts oscillator.

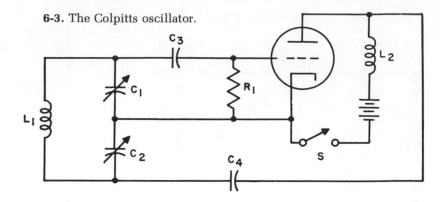

high impedance to RF it forces the RF component of the plate current to flow through C_4 to C_2, where it is wanted.

The Tuned-Grid Tuned-Plate Oscillator

As its name implies, this circuit (Fig. 6-4) has one tuned circuit L_1/C_1 in its grid and another L_2/C_3 in its plate. When the plate is tuned to a slightly higher frequency than the grid, it tends to return energy to the latter through the capacitance formed by the actual plate and grid elements in the tube. This interelectrode capacitance is represented by the dotted lines

6-4. The tuned-grid tuned-plate oscillator.

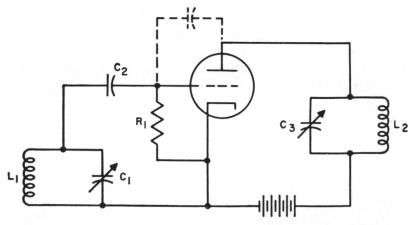

in the diagram. The frequency of oscillation is determined by
the L-C circuit with the lower Q. This symbol stands for *quality*, and is the ratio of the reactance of either L or C to the
series resistance. In practical circuits the tuning capacitor,
with its air insulation, has virtually infinite resistance, so the
determining Q factor is the small but appreciable resistance of
the tuning coil.

Known as TGTP for short, this type of circuit requires two
completely independent L-C circuits, not inductively coupled,
and is tricky to adjust. It was used in the early days of shortwave transmission when oscillator stability wasn't as important as it is now, and is mentioned here only because it is the
basis for the crystal-controlled oscillator.

Crystal-Controlled Oscillators

Small, flat pieces of quartz crystal have the extraordinary
property of acting exactly like conventional combinations of
inductance and capacitance, with the improvement that they
have extremely high Q. The frequency of oscillation is determined by their thickness, and can be adjusted to very close
tolerances merely by grinding. The thinner the crystal, the

6-5. (A) Two crystal holders and (B) a disassembled crystal.

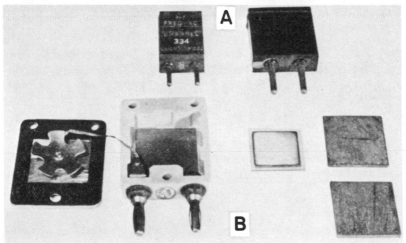

higher the frequency. Actual crystals used in communications equipment are about the size of ordinary postage stamps and are protected in simple holders with brass plates making contact with the parallel surfaces (Fig. 6-5).

Known as *rocks* in ham lingo, crystals are inexpensive, reliable, accurate, and foolproof, and therefore very popular with amateurs.

The Crystal-Controlled TGTP Oscillator

If the L_1/C_1 circuit of Fig. 6-4 is replaced by a crystal, as in Fig. 6-6, we still have a TGTP oscillator, but with only one tunable circuit, L/C. The adjustment of the latter is not critical, the frequency now being determined entirely by the very high Q crystal. As simple as this circuit is, it is entirely dependable.

The Pierce Crystal Oscillator

The Pierce crystal oscillator (Fig. 6-7) is actually a Colpitts oscillator in which the tuned circuit is replaced by a crystal. Voltage division is accomplished through the plate-cathode and grid-cathode interelectrode capacitance of the tube. The Pierce circuit is widely used because it is even simpler than the TGTP crystal-controlled oscillator.

The Single-Tube Transmitter

It is possible to feed the output of a simple oscillator directly to an antenna and to accomplish fairly good CW communication with the combination. However, this has certain drawbacks. Because tubes and circuit components are cheap, most amateurs (even beginners) use one or more additional stages between the oscillator and the aerial. These stages not

6-6. A circuit of a basic crystal-controlled oscillator.

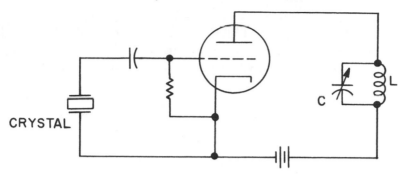

6-7. (A) The equivalent circuit of a Pierce crystal oscillator is similar to that of a Colpitts; (B) a practical Pierce circuit.

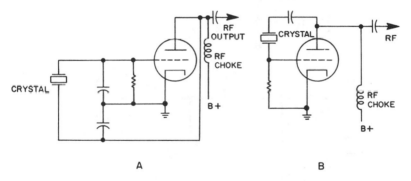

A   B

only produce a stronger output signal but also permit more flexible control of the radiated carrier frequency.

Interstage Coupling

The coupling between the various stages of a transmitter using one or more stages of amplification after the oscillator should transfer the radio signal with as little energy loss as possible. There are two basic methods of interstage coupling, *capacitive coupling* and *inductive coupling*.

Figure 6-8 illustrates two methods of capacitive coupling between RF stages. In (A), V_1 might be the oscillator tube, and L_1/C_1 its plate-tuning elements. The coupling to the amplifier tube V_2 is provided by capacitor C_2, which may be either fixed

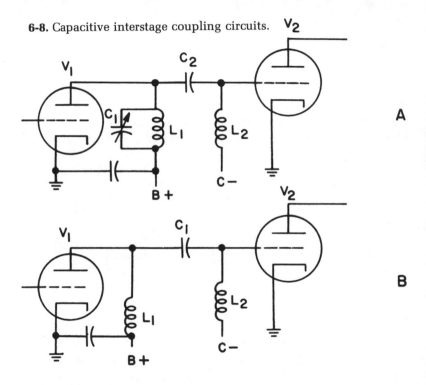

6-8. Capacitive interstage coupling circuits.

or variable. If it is variable, it acts to some extent to control the amplitude of the signal fed from V_1 to V_2. In the grid circuit of the latter, L_2 is merely an RF choke that permits grid bias to come through from the bias source without letting RF energy short-circuit itself to that source.

In circuit (B), V_1 might be an intermediate stage and V_2 a succeeding one. The plate circuit of V_1 here is not tuned, but its DC circuit is completed by the choke L_1, which forces the amplified signal at the plate to go through coupling capacitor

C_1 to the grid of V_2. The L_2 again is an RF choke to keep the amplified signal going to the grid.

Inductive coupling also takes two general forms. Figure 6-9 shows what might well be the IF stage of a superheterodyne receiver. The T_1 is a transformer, with the tuned primary in the plate circuit of the driver tube V_1 and the tuned secondary in the grid circuit of the driven stage V_2. In some low-powered transmitters T_1 is actually a regulation IF transformer.

Sometimes it is desirable for purposes of safety, especially when high voltages are involved, to have complete physical separation between one stage and another. This is done by link coupling, as shown in Fig. 6-9. In the plate circuit of V_1, coils L_1 and L_{1A} are primary and secondary, respectively, of an RF transformer, with L_1 tuned by C_1 in the usual manner. In the grid circuit of V_2, L_2 and L_{2B} are secondary and primary of an identical transformer. The L_1 induces energy in L_{1A}; this travels through the connecting link between L_{1A} and L_{2A}, and

6-9. Inductive coupling circuits.

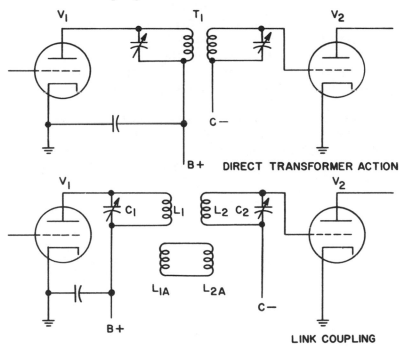

L_{24} induces the energy into L_2. The link itself can be twisted-pair wire or coaxial cable. Because of the dual tuning circuits, link-coupled stages require very careful adjustment for best results.

RF Power Amplification

Radio-frequency power amplifiers are operated class C, which means that the grid is biased slightly below cutoff so that plate current flows for less than 180 degrees of the input cycle. Operating an amplifier class C assures high plate efficiency and permits more than normal power to be applied to the tube without exceeding its ability to handle the increased current flow.* In other words, since the tube is nonconducting or resting for less than half of the input cycle, it can pass more current when it does conduct without being damaged. The fact that an RF amplifier operates class C also explains the need for a tuned *tank* circuit as a plate or output load. If a nonresonant plate load were used, the output signal would be an amplified version of that positive portion of the input signal during which the tube conducts. Thus, assuming that the input signal is a sine wave, the output signal would be something less than half a sine wave. However, if a resonant tank circuit is used as a plate load for a class C amplifier, the output signal across the tank is always a sine wave at the frequency of L-C resonance. One way to look at it is to imagine the "flywheel" action of a tank circuit (or its inherent nature to oscillate at its resonant frequency) as filling in the portion of the input sine-wave signal cut off when the tube is nonconducting.

The output impedance or plate load of an RF power-

---

* The output power of a transmitter using a class C final amplifier may be calculated as approximately 60 percent of the input power to the final amplifier. The input power to a tube is determined as plate current times plate voltage.

amplifier tube is adjusted by tuning the plate tank circuit to resonance so that it effectively acts as a pure resistance at the operating frequency. Increasing the amount of energy transferred to the grid of a following amplifier (or in the case of the final amplifier, an antenna) effectively reduces the plate load resistance (the DC resistance of the resonant tank circuit) of the driving tube. Since reducing the plate resistance of an amplifier increases the plate current, we can see how adjusting the coupling or "loading" of the plate tank circuit of a power-amplifier stage affects tube operation. For example, it is usually desirable for the final RF power amplifier of a transmitter to deliver maximum power to the antenna. To do this, coupling between the antenna and the plate tank circuit is increased until the tube draws its maximum rated plate current. However, when increasing the coupling to obtain the desired plate current, care must be taken to keep the plate tank circuit itself adjusted to resonance.

When an RF amplifier has input and output circuits tuned to the same frequency, it oscillates like a tuned-grid tuned-plate oscillator, unless steps are taken to lessen the effect of feedback through the grid-to-plate interelectrode capacitance. In pentodes and tetrodes this grid-to-plate capacitance is reduced enough by the internal shielding of the screen grid, and no special circuits are required. However, since tetrodes and pentodes tend to oscillate with very small values of feedback voltages, care must be exorcised to prevent the possibility of external feedback. This requires good isolation between plate and grid circuits.

Neutralization

When a triode is used, a special circuit must be used to reduce the feedback through the grid-to-plate capacitance, since it has no screen grid to shield the grid from plate. This circuit accomplishes what is known as *neutralization*. Essentially, neutralization is accomplished by taking a portion of the RF current from either the grid or plate circuit and applying it to the other circuit, so that it effectively cancels the current flowing between grid and plate because of the interelectrode capacitance. For complete neutralization, the two RF currents must be equal in amplitude and 180 degrees out of phase. One method of neutralizing, known as *plate neutralization*, is shown in Fig. 6-10. The circuit shown here uses a balanced output with voltage division obtained by the split-tuning capacitor of the tank circuit. This means that the voltage at the top of the tank-circuit coil and at the plate is 180 degrees out of phase with the voltage at the bottom of the tank-circuit coil. The capacitor C_N, the neutralizing capacitor, thus picks off a portion of the RF signal current from the bottom of the tank-circuit coil and applies it to the grid to cancel the RF signal fed back because of the plate-to-grid interelectrode capacitance. To achieve canceling with a voltage of equal amplitude, C_N is adjusted until its capacitance is approximately equal to the plate-to-grid capacitance of the tube. The actual process of neutralizing requires a milliammeter to read the rectified DC grid current. If a triode RF amplifier is not neutralized, tuning the plate circuit through resonance causes a perceptible drop in grid current. The neutralizing capacitor is adjusted so that tuning the plate circuit through resonance has no pronounced effect on grid current.

Frequency Multiplication

In some transmitters it is desirable to have the transmitted frequency a multiple of the oscillator frequency. This is accomplished by tuning the RF amplifier following the oscillator to a *harmonic* or multiple of the oscillator frequency. Theoretically, the plate tank circuit can be tuned to any harmonic

---

* The fundamental frequency is called the first harmonic. For example, 200 hertz is the second harmonic of 100 hertz, 300 hertz the third harmonic, and so forth.

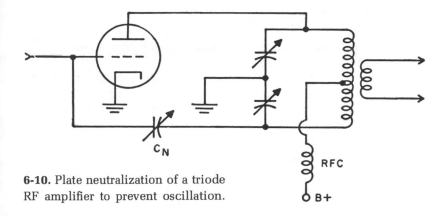

6-10. Plate neutralization of a triode
RF amplifier to prevent oscillation.

needs to be neutralized. Since the plate and grid circuits are tuned to different frequencies, there is no danger of oscillation due to grid-to-plate interelectrode capacitance. Using two or more doublers in succession permits obtaining an output frequency that is 4, 8, 16, etc., times the original oscillator frequency.

Keying

Keying a transmitter is nothing more than turning it on and off by means of a switch, the telegraph key. Methods of keying are referred to by such names as blocked-grid keying, plate-circuit keying, and cathode keying.

Blocked-grid keying works by applying a blocking bias to the control or suppressor grid of a transmitter's oscillator or amplifier tube when the key is open. This effectively shuts off a transmitter, since the tube which is biased to cutoff ceases to conduct. When the key is closed, the bias is removed and the tube conducts to "turn on" the transmitter. With grid-blocked of the input signal.* However, since there is a marked drop in plate efficiency beyond the second harmonic, most multipliers are usually doublers, with the plate tank tuned to the second harmonic of the grid signal. A multiplier using a triode never keying, because it handles a small current flow, there is little chance for the key to spark.

Plate-circuit keying, which works by turning on and off the power to the plate of a transmitter's stage, is effective since it guarantees that the keyed stage will not conduct when the key is open. For plate-circuit keying, a key is placed in series with the negative lead from the plate power supply. A key also may be effectively placed in series with the positive lead; however, this requires a keying relay to isolate the key, since this point is a substantial voltage above ground. Plate keying is not used very much.

Cathode keying works by opening and closing the cathode lead to a tube. This effectively opens and closes the DC circuits of both the plate and grid at the same time. Cathode keying is very similar to plate-circuit keying, although it usually produces less arcing at the key contacts than plate-circuit keying.

When a transmitter is not completely turned off between the dots and dashes, as in the case of improper blocked-grid keying, it produces what is known as a *backwave*. When a transmitter generating backwave radiations is keyed, the transmitted dots and dashes heard at a receiver would be merely louder portions of a continuous tone. This makes the code difficult to copy. A backwave may also be produced by keying a triode final amplifier that has not been properly neutralized. Backwave radiation is easily discovered by monitoring your own transmission.

The wave shape of CW transmission is also interesting. Theoretically perfect keying would produce the square envelopes shown in Fig. 6-11(A). Here, turning the transmitter on and off is accomplished instantaneously. Actually, this perfect keying is not desirable, because the resulting square envelope contains an infinite number of harmonic frequencies that produce short pulses of energy throughout the entire radio spectrum. These very short pulses of wide-band energy are known as *key clicks*. Although it is possible to key a transmitter so as to produce key clicks at both the beginning and end of a code

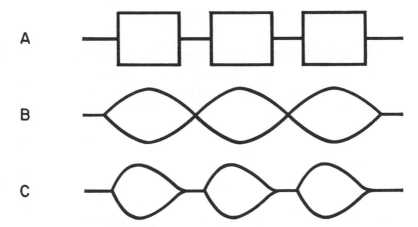

6-11. Keying envelopes: (A) theoretically perfect keying, (B) completely sinusoidal keying with a slow buildup and decay, and (C) desirable keying with a faster buildup than decay.

character, the selective nature of the tuned circuit eliminates the higher harmonics. Thus, a transmitter's tuned circuits tend to round off the keying envelope to produce one with a more gradual rise and fall from zero to maximum transmitted energy. Figure 6-11 shows different keying envelopes: (A) and (B) represent the two extremes of completely rectangular and sinusoidal envelopes; (C) shows desirable keying. Notice that a desirable envelope has a faster buildup than decay. This relatively fast buildup at the beginning of a code character produces a slight key click that makes the characters more pleasing to hear and easier to read, yet does not interfere with transmission on nearby frequencies.

Voice Modulation

A discussion of voice modulation necessarily begins with the actual sound or vibratory disturbance of air that is set up when a person talks. Sound waves can vary over a relatively wide frequency range. The range of sound that humans can hear (the audible range) varies from around 15 to 18,000 hertz. The vibratory nature of sound makes it relatively simple to modulate an RF carrier with sound intelligence even though plotted wave forms of most sounds, such as the human voice, are complex combinations of a fundamental sine wave and its harmonics. First of all, sound such as the spoken word is transformed into an AC voltage varying at the same frequency and relative amplitude as the sound vibrations. This is accomplished by a microphone. The AC voltages representing sound intelligence at the output of a microphone are then amplified to a strength sufficient for modulating (varying the amplitude of) a carrier current before it is applied to a transmitter's antenna. Alternating-current voltages representative of sound vibrations are referred to as *audio voltages*. Vacuum-tube or transistor stages that amplify audio voltages are referred to as *audio amplifiers*.

The process of communicating by radio is completed when a receiver picks up the modulated radio waves, amplifies them, extracts the audio voltages, and finally applies these audio voltages to a speaker or earphones which do the reverse of the microphone and convert them back into the original sound intelligence.

Amplitude Modulation

There are certain conditions that affect the "pureness" as well as the intensity of the modulated intelligence. For example, it is important that the amplitude of the RF carrier be constant before being modulated. Poor filtering of the transmitter's power supply could impose a 60-hertz variation on the RF carrier, which would result in a 60-hertz hum being heard by any receiver picking up the transmission.

The *percentage of modulation*, or the depth to which an RF carrier is modulated, determines the strength of the audible output. The maximum occurs with 100 percent modulation, or when the carrier is at intervals reduced to zero as well as

raised to a peak amplitude of twice its unmodulated amplitude. Four degrees of modulation are illustrated in Fig. 6-12:
(A) shows an unmodulated carrier with an amplitude A; (B)

6-12. Percentage of modulation: (A) unmodulated carrier, (B) 50 percent, (C) 100 percent, and (D) overmodulation.

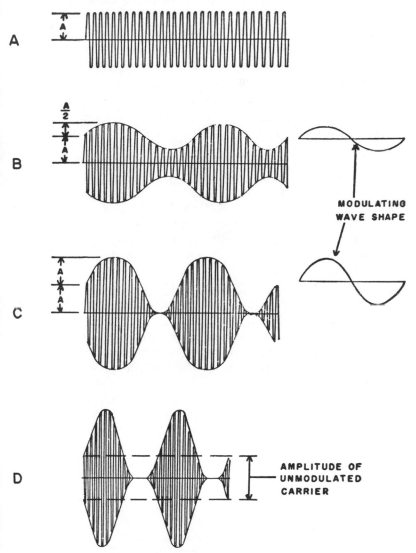

shows 50 percent modulation, the modulation causing a maximum increase and decrease of one-half the carrier's unmodulated amplitude, or $A/2$, and (C) shows 100 percent modulation, the modulation causing a maximum increase and decrease equal to the unmodulated amplitude of the carrier, or A. Notice that for 100 percent modulation the carrier is reduced to zero as well as increased to twice its unmodulated amplitude, or a total of 2A. Figure 6-12(D) shows the effect of overmodulation. Here the peak amplitude exceeds twice the unmodulated carrier amplitude, and the shape of the modulating wave has been distorted. Consequently, overmodulation distorts the actual intelligence.

Sidebands

When we modulate an RF carrier with audio, we combine two frequencies—the carrier frequency and, at any given instant, the frequency of the audio signal.

Two actions now occur. First, the hitherto *continuous*-wave carrier varies up and down in amplitude in direct accord with the original voice variations; hence the name *amplitude modulation* (AM) for this method. Second, the mixing action of the two separate RF and AF currents causes the carrier to spread out in frequency by an amount equal to the *sum* of the frequencies in one direction and their *difference* in the other. These shifts are called *sidebands*.

For example, let us assume a typical amateur carrier of 3,900 kHz and an AF modulation signal of 2 kHz. In addition to the existing carrier, we now get 3,900 plus 2, or 3,902 kHz, which is called the *upper sideband,* and 3,900 minus 2, or 3,898 kHz, the *lower sideband.* The total signal that goes out over the air is no longer strictly limited to 3,900 kHz but is 4 kHz wide with 3,900 as the center point. The same audio modulation appears in both sidebands, so what we have in effect is two different RF signals with the same AM signal on them.

Single Sideband During the early days of radiotelephony it occurred to engineers that either sideband, and the carrier as well, could be eliminated from the transmission without affecting the basic voice intelligence contained in the remaining sideband. The system that accomplishes this reduction is called *single-sideband suppressed carrier*, invariably shortened to *sideband* or SSB. Initially, the equipment needed to cancel out two-thirds of the AM signal was both complicated and expensive, so the method was used only on commercial radiotelephone circuits. The situation changed after World War II with the development of improved equipment and techniques not only by manufacturers but by radio amateurs themselves. SSB has now largely replaced conventional double sideband AM on the ham frequencies up to 30 megahertz; in fact, there is no provision at all for the latter method in most of the multiband ham receivers, transmitters, and transceivers now on the market. In most sideband equipment the operator has a choice of using either the upper sideband, usually abbreviated to USB, or the lower, LSB.

It should be understood that SSB is a different method of transmission, not of modulation. What are its advantages over the older method? Briefly, they are:

1. Two stations can share the same carrier frequency, *without interference,* if one is on upper sideband and the other is on lower sideband. *This is actually equivalent to doubling the width of the ham bands!*

2. For the same transmitter power, SSB puts out a more potent voice signal.

3. SSB speech is extremely "clean," in the sense that it is free of spurious signals that can wipe out sound or sight, or both, in neighbors' television or radio receivers. Without SSB, hams could not possibly survive in apartment houses or other crowded residential facilities.

Improvements over the years notwithstanding, SSB circuitry is still quite complex and cannot be explained in simple terms. Don't let this keep you from enjoying its features. After you've used an SSB rig for a while and are throughly familiar with its controls and general operation, you might want to read yourself into the theory of its phase-locked loop, balanced modulator, radio-frequency processor, marker board, digital hold, anti-VOX amplifier, etc. There's a lot to learn! Some excellent texts on the subject are listed in the appendix.

Frequency Modulation

To owners of high-grade stereo tuners and amplifiers, the letters FM usually are interpreted as *fine music,* but of course this is only a popular version. To the technically-minded it means *frequency modulation,* a system quite different from AM. In the latter, the carrier-wave frequency remains fixed and the amplitude of the signal bounces up and down with the variations of the speech signals imposed on it. In FM, the amplitude of the carrier is fixed, but its frequency varies back and forth with the speech.

The advantages of FM are two-fold:

1. It is virtually immune to the kinds of noise and interference that often plague AM reception, because these are basically AM in nature and do not readily get through an FM receiver's circuitry.

2. An FM receiver can easily be *squelched.* This is done by raising the bias on its early-stage amplifiers so that only signals above a particular level can get through, while weaker ones don't get through at all. In the absence of a desired signal, a properly-adjusted set doesn't make a sound, but when the signal does arrive it blasts out of the loudspeaker in great style.

With weak signals discouraged by the squelch action (more properly, this is called *limiting*), the chances of working DX (distance) on FM are obviously rather slim. This is a small disadvantage, because FM is used only on high frequencies that generally are only suitable anyway for line-of-sight communication. If you want to hear weak stations, merely turn down the squelch knob a bit. For more on FM, see Chapter 14.

Radio-Wave Propagation

Radio waves travel with the speed of light and can be reflected from various layers of the earth's atmosphere and from the earth itself. They travel not only along the surface of the earth but also through the upper atmosphere. The part of the wave or energy from the antenna that travels along the surface of the earth is called the *ground wave;* the part that goes out at an angle above the horizontal is called the *sky wave* or *ionospheric wave.* (The ionosphere is the region of rarefied and ionized atmosphere surrounding the earth at a distance of from 50 to 200 miles.) A third wave, called the *tropospheric wave,* is that part of the original wave which is refracted and reflected in the troposphere, an area of clouds and storms from 3 to 7 miles high.

The importance of the ionosphere (also called the Kennelly-Heaviside layer) in the propagation and transmission of radio signals is now well recognized. If it were not for the existence of this layer much of the energy emitted by a shortwave transmitter would escape into space and be lost.

It may seem at first that the sky waves would be lost and therefore of no value in communication. However, as they leave the transmitting antenna they travel until they meet one of the ionized layers above the earth. Striking one of these regions of ionized particles, the wave is bent or reflected back toward the earth in much the same manner as a light ray striking a mirror. Thus, instead of passing off into space, the signal eventually comes back to earth at a point hundreds and sometimes thousands of miles from the starting place. The distance between the transmitter and the return point is called the *skip distance* and comprises the area over which the station cannot be heard.

The approximate positions of the principal ionized layers are illustrated in Fig. 6-13. Tests have shown that most of the waves transmitted at night pass through the E layer and on to one of the F layers before they are reflected back to earth. During daytime, each of the three layers' reflective properties is dependent on the frequency of the waves.

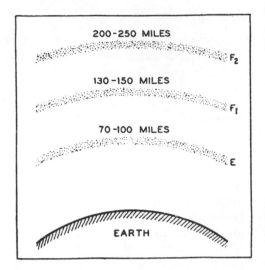

6-13. The location of the principal ionized layers affecting the transmission of radio waves at certain frequencies.

The F_2 or highest layer of the ionosphere is responsible for most of the long-distance radio contacts in high-frequency communication below 28,000 kHz. The efficiency of this frequency range varies according to a well-defined cycle or system of cycles, related to the eleven-year solar cycle as well as to daily and seasonal variations of the sun with respect to the earth.

The E layer is often responsible for reflections of signals above 28 MHz. This so-called *short skip* or *sporadic E skip* is quite unpredictable as to season or time of day, probably because of irregularities of ionization in the E layer.

7

Electronic Components

Conductors of Electricity

About 95 percent of the conductors in electronic equipment are made from copper. This metal is highly ductile, and can be drawn into wire so fine as to be almost invisible. It is a very good conductor; that is, its resistance to the movement of electrons is low. Of the common metals only silver is better, and then only slightly. By comparison with copper, brass is 4 times more resistive, aluminum 1.6, iron 5.6, and tin 6.7. An extra feature of copper is the ease with which it can be soldered.

The other 5 percent represent alloys of metals deliberately formulated to have high resistance, for the control of current in numerous circuits or to produce heat.

For some unknown reason, wire sizes are indicated by gauge numbers in inverted order: the smaller the number the thicker the wire, and the larger the number the thinner the wire. The numbers run from 1 through 40, but there are no standard wires with odd gauge numbers because the difference between one gauge and the next one up or down is only a few thousandths of an inch. A typical listing in a radio catalog shows wires in sizes 16, 18, 20, 22, and 24, but no 15, 17, 19, 21, or 23.

The most common type of wire used for connecting the various parts of radio equipment is referred to as *hookup wire*. It employs either a solid or a stranded conductor covered by cloth or plastic insulation. A stranded conductor, made by twisting many small wires together, has the advantage of being more flexible and harder to break by bending than a solid conductor. Radio hookup wire is usually #20 gauge solid. The standard for line cords, which connect equipment to the power supply, is #16 or #18 stranded.

In addition to hookup wire, there are many special types of conductors such as *high-voltage cable, multiconductor cable,* and *shielded cable.* High-voltage cable uses a small conductor surrounded by thick insulation, which prevents the high voltage from arcing to ground. Small wire is adequate because in most high-voltage applications (2,000 volts or more) the current flow is small. Multiconductor cable incorporates many insulated single-conductor wires in a protective casing. The individual conductors are marked for identification by coverings of different colors. Shielded cable consists of one or more insulated conductors surrounded by a flexible copper braid. Relatively thin single-conductor shielded cable, about ⅛ inch in diameter, is commonly used for connecting microphones to transmitters. Larger cable, from about ¼ inch to ½ inch, is called *coaxial cable* or more simply *coax,* and is used between antennas and transmitting/receiving equipment. See Fig. 7-1.

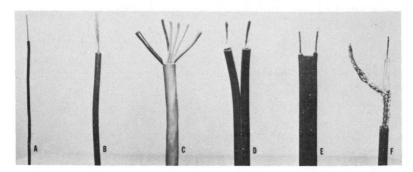

7-1. Some of the common conductors used in radio equipment: (A) solid single-conductor wire, (B) stranded wire, (C) multiconductor cable, almost always with stranded wire, (D) two-wire "zip" cord, for connection to power lines, (E) antenna "twin-lead," (F) shielded single-conductor wire, usually known as "coaxial cable" or "coax."

Insulators

An insulator is any nonconducting material used to isolate one conductor from another or from a radio set's chassis, which is invariably grounded. Insulation of some sort is an integral part of virtually all electronic components. It takes the form of paper, mica, glass, fabric, fiber, rubber, wood, enamel, ceramic, etc., and of a very large variety of plastics that can be molded or machined. Insulators for outdoor use, specifically for supporting antennas and their related wires, are invariably made of smooth-surfaced glass, ceramic, or plastic, because these materials shed water readily.

All insulation, indoor and outdoor, should be kept as clean and dry as circumstances permit. Antennas mounted on or near chimneys are particularly vulnerable to furnace soot, which is quite a good conductive material; their insulators should therefore be inspected regularly and replaced if they cannot be cleaned.

Resistors

The value of resistors is expressed in ohms and their power-handling capability by their wattage ratings, while the effect that they have on current and voltage in electric circuits may be calculated by using Ohm's law. *Insulated* or *uninsulated carbon resistors* are the most frequently used. They are inexpensive, moderately accurate, and come in a very wide range of resistance values and wattage ratings. They are fitted with tinned wire leads about 2 inches long, usually called *pigtails*. See Figs. 7-2, 7-3, and 7-4.

The value of fixed carbon resistors can be determined from their color markings, which follow a code used by all manufacturers. On uninsulated resistors (Fig. 7-3), the color of the body indicates the first digit of the resistance value; the color on the end indicates the second digit; and the color of the dot tells the number of zeros to add to the first and second digits to obtain the total resistance. On insulated resistors, the colors

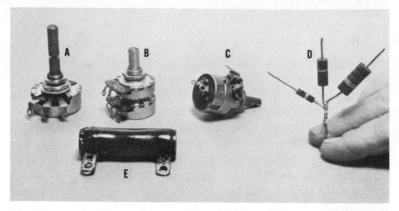

7-2. Representative resistors: (A) single potentiometer, (B) dual "pot," (C) volume-control pot with an attached switch, (D) ½-, 1-, and 2-watt carbon resistors, (E) wire-wound resistor with ceramic coating.

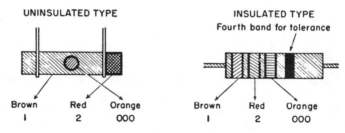

UNINSULATED TYPE

INSULATED TYPE
Fourth band for tolerance

Brown	Red	Orange
I	2	OOO

Brown	Red	Orange
I	2	OOO

7-3. Color code for carbon resistors. (*Courtesy of the Heath Company.*)

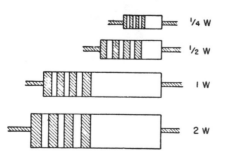

¼ W

½ W

I W

2 W

7-4. Relative full size and corresponding wattage ratings of carbon resistors. (*Courtesy of the Heath Company.*)

are a series of bands (Fig. 7-3). The first band indicates the first digit, the second band the second digit, and the third band the number of zeros to add to the first and second digits. The numerical values assigned to the various colors used in the code are given in the following table.

COLOR CODE FOR RESISTORS

Uninsulated→ or Insulated→ color	Body First Ring (*first figure*)	End Second Ring (*second figure*)	Dot Third Ring (*third figure*)
Black	0	0	None
Brown	1	1	0
Red	2	2	00
Orange	3	3	000
Yellow	4	4	0,000
Green	5	5	00,000
Blue	6	6	000,000
Violet	7	7	0,000,000
Gray	8	8	00,000,000
White	9	9	000,000,000

We can see from the table that the value of both resistors shown in Fig. 7-3 is 12,000 ohms. In insulated carbon resistors, a fourth band is sometimes added to indicate the tolerance of the specified value of resistance. When no band is used, the resistance can be expected to be within ±20 percent of the specified value; silver indicates a ±10 percent and gold indicates ±5 percent tolerance.

Figure 7-4 indicates the relative size of carbon resistors and their corresponding wattage ratings.

It is often difficult to identify the narrow color bands on ¼-watt resistors, the smallest size made and the one found in the greatest number in amateur equipment. Browns and reds as one pair, and blues and yellows as another, sometimes cannot be told apart, especially under fluorescent lighting. Whenever there is the slightest doubt about band colors, it is advisable to measure the actual resistance of suspected units. Fortunately, this is a very quick and easy job with a volt-ohmmeter or digital readout meter such as the ones described in Chapter 9.

Wire-wound resistors (Fig. 7-2) are made by winding high-resistance wire around a form so that each turn is insulated from adjacent turns. They cost more than carbon resistors, but they have closer tolerances for their specified values of resistance. Wire-wound resistors with high wattage ratings can also be made relatively small. For these reasons, wire-wound resistors are used where accurate resistance or high current flow is required.

Adjustable wire-wound resistors have part of the winding left bare so that a sliding metal band can tap off any amount of resistance between zero and maximum. Wire-wound resistors can also have any reasonable number of fixed taps. Wire-wound resistors with fixed or variable taps are also called *voltage dividers*. The resistance values and wattage ratings of wire-wound resistors are usually printed directly on their bodies.

Variable resistors, as distinguished from adjustable ones, have a ring-shaped resistance element of carbon or wire, over which a shaft-controlled contact arm passes. They are available in many sizes. If connection is made only to one end of the element and to the contact, the device is called a *rheostat.* If both ends and the contact are used, it becomes a *potentiometer* (*pot*, for short). The distinction lies in circuit application rather than physical construction; obviously, a potentiometer becomes a rheostat if one terminal of the resistance element is merely left idle. A rheostat is always connected in series with a circuit, and acts to limit the flow of current in it. A pot is usually connected across a circuit and functions as a voltage divider; that's what the name implies—potential (i.e., voltage) regulator.

Turning the shaft moves the contact over the resistive element to give any resistance value from zero to maximum. Wire-wound are more expensive than carbon potentiometers and are used for accurate control or high current flow.

Switches to control the AC power line or other circuit elements are often mounted on the backs of pots. In receivers the volume control is the favorite. These switches are turned on and off either by a push-pull action of the central shaft or by the first few degrees of the latter's rotation near the seven o'clock setting of the knob.

Transformers

When two or more coils are inductively coupled, an AC voltage applied to one coil will cause an AC voltage of the same frequency but opposite polarity to be induced in the adjacent coil or coils. Such an arrangement is called a *transformer*.

For use on audio frequencies, which means everything from about 60 hertz house current up through speech and music, practically all transformers have box-shaped, laminated sheet-iron cores, over which the coils are wound. See Fig. 7-5. For use on radio frequencies, transformers have either air cores,

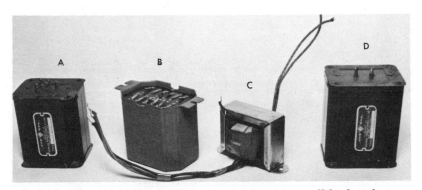

7-5. Some representative low-frequency units: (A) small high-voltage power transformer, (B) output transformer with multitapped primary and secondary, (C) open-type power transformer with wire leads instead of terminals posts, and (D) filter choke.

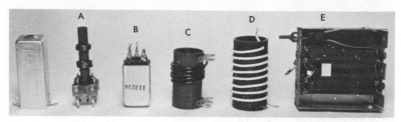

7-6. Radio-frequency units: (A) IF transformer removed from its aluminum case to show two separated windings on fiber form, (B) cased RF transformer, showing terminal lugs in bottom, (C) close-wound tuning coil using heavy enameled wire, (D) spaced tuning coil of heavy bare wire, and (E) double tuning coil of widely spaced wire on ribbed form.

which is to say no cores at all, or small cores of finely powdered iron in the form of pressed sticks. See Fig 7-6. In the air-core type, windings are supported physically on insulating tubing. The powdered-iron type is similar, with the addition of the core itself. This is a little slug that looks like a blackened 10-24 or ¼-20 screw; in fact, in some transformers it is threaded and it can be adjusted axially inside the form on which the coils themselves are wound. In other transformers the slug is attached to a small brass rod, the end of which protrudes through the case and is slotted for convenient screwdriver adjustment.

It is not absolutely necessary for a transformer to have two windings insulated from each other; it can have a *single* winding and still work like a transformer. The explanation of this seeming anomaly lies in Fig. 7-7. The single winding has a tap on it, anywhere along its length. If the transformer is intended for stepping down voltages, the entire coil, section *A*, functions as the primary; the part between the tap and the bottom end, section *B*, works as the secondary. If step-up action is desired, *B* is simply used as the primary and *A* as the secondary. Such a device is called an *autotransformer*, and is useful for numerous AF and RF applications.

It is important to remember that all transformers work on

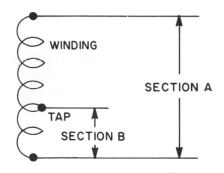

7-7. In the autotransformer, all of the winding, section A, can be either primary or secondary, and the tapped section B can be either secondary or primary.

the same principle. The difference between AF and RF types is mainly that the former handle heavy currents, the latter very small currents.

An important feature of transformers is that the voltage induced in the secondary coils can be either greater or smaller than the applied voltage in the primary coil, depending upon the number of turns in the respective coils. For example, assume that we have a perfect transformer in which all of the magnetic lines of force generated in the primary coil are coupled to a secondary coil. Then, if a primary winding of 200 turns has 200 volts applied to it, the magnetic flux established will induce 1 volt in each turn of the secondary coil or winding. Thus, when the secondary winding consists of 100 turns, the total voltage across it will be 100 volts. This is a reduction from the 200 volts on the primary to 100 volts on the secondary winding, or a step-down of 2 to 1. On the other hand, if the secondary winding has 2,000 turns, each with 1 volt of induced potential, the voltage across the winding will be 2,000 volts. In this case, the voltage has been stepped up from 200 to 2,000 volts, or by a 1 to 10 step-up ratio.

A component that looks and feels like a transformer but shows only two connections is a *filter choke*. It has a single untapped winding on a laminated iron core, as in a transformer. Its purpose is to help smooth out the variations of rectified current in an AC power pack feeding a receiver or a transmitter. In some applications two chokes are used in series and are supplemented by one, two, or three high-value capacitors. In tube-type equipment the AC power transformer and the accompanying filter system can easily outweigh the receiver or transmitter itself. The power needs of solid-state ham sets are much lower, and in many cases the entire pack can be built right into the main chassis.

Single-Winding Inductors

Single-winding inductors serve a number of purposes, depending on their number of turns and core material. See Figs. 7-6 and 7-8. Those having relatively small windings, say a few turns to a couple of hundred turns, with air or powdered-iron cores, can be used with variable capacitors to form resonant circuits at radio frequencies. In this application they are called *tuning coils*. Identical or similar inductors can be used to control the passage of RF currents in various circuits. In this application they are called RF *choke coils*, or more commonly, RF *chokes*. Because of the wide frequency bands open to amateurs, these RF coils and chokes vary very widely in size and shape.

Many of the popular sizes of RF chokes consist of a number of small, flat sections of wire, called *pies*, separated about an eighth of an inch on a ceramic form. The wire itself is wound in a honeycomb pattern, so that the turns are slightly separated. The purpose of both spacings is to minimize the capacitance action between adjacent wires and groups of wires. This must be taken into consideration because it has the unwanted effect of turning the choke into a resonant circuit all by itself or of nullifying its choking action by providing a low-impedance path of its own to the very RF currents the choke is supposed to block off.

Capacitors

Capacitors with fixed and variable amounts of capacitance are extensively used in radio. See Figs. 7-9 and 7-10. Fixed capa-

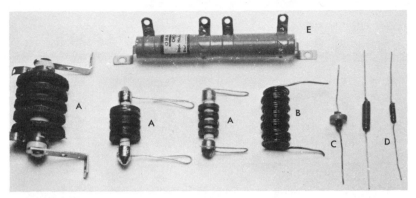

7-8. Radio-frequency chokes widely used in receivers and transmitters: (A) three sizes of pie-wound type, (B) jumble-wound type of heavy wire, (C) iron-core type, having high inductance in small package, (D) two sizes of small single-layer type, and (E) heavy-duty line choke, two units on common form.

citors perform such services as filtering the outputs of power supplies and preventing the flow of direct current in circuits where it would conflict with AC in the same path. Variable capacitors are used mostly in series or parallel-resonant circuits with RF coils. Since the frequency of these circuits is dependent upon the values of L and C, varying the capacitance is one easy way of varying the resonant frequency.

As the unit of capacitance, the farad is much too large for practical purposes. The *microfarad*, or one-millionth of a farad, is more commonly used instead. This is abbreviated to μf, μ being the Greek letter *mu*, pronounced "me-you." Frequently, the abbreviation is shown as *mf* or *MF* because the printer does not have Greek letters available in his composing room.

Very small capacitors are rated in *micromicrofarads* ($\mu\mu f$) or by a newer term, the *picofarad* (*pf*), which means the same. The shift to picofarad is taking place gradually, and both terms will be encountered in electronic practice probably for many years, or until manufacturers use up their old labels and stamps.

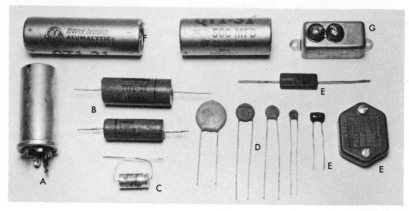

7-9. Fixed capacitors: (A) aluminum-can electrolytic, twist-lock mounting, (B) two sizes of paper type, (C) small low-voltage electrolytic, (D) four sizes of ceramic disks, (E) three types of mica, (F) two sizes of paper-cased electrolytics, and (G) "bathtub"-style paper.

7-10. Common variable capacitors: (A) small screwdriver-adjustable trimmers, (B) close-spaced small variable, (C) three sizes of wide-spaced transmitting variables, and (D) close-spaced receiving type.

Mica capacitors, made with conducting surfaces separated by thin sheets of mica, have values ranging from about 1 pf to 0.01 μf. They also have very low losses. The capacitance of a mica capacitor is often expressed by a color code on the body. The numerical values assigned to the various colors are the same as for carbon resistors. There are three different methods of placing the colored dots used to code mica capacitors (Fig. 7-11). If there is only one row of dots, they are read in the di-

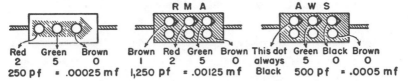

Red   Green   Brown   Brown Red Green Brown This dot Green Black Brown
2   5   0   1   2   5   0   always   5   0   0
250 pf = .00025 mf 1,250 pf = .00125 mf Black 500 pf = .0005 mf

7-11. Color code for mica capacitors. Note alternate abbreviations "pf" and "mf" (μf) for picofarad and microfarad. (*Courtesy of the Heath Company.*)

rection of the arrow that appears on the capacitor as shown. The first color gives the first digit of the rating in pf, the second color the second digit, and the third color the third digit. If there are two rows of dots, the code used may be either of two types, the RMA (Radio Manufacturers' Association) or AWS (American War Standard), although the latter type should be appearing less frequently now. In both codes the first three dots, read in the direction of the arrows, indicate the first three digits of the pf rating, while the third dot on the bottom row gives the decimal multiplier. In the RMA code, the first two dots in the bottom row indicate, respectively, the voltage rating and tolerance, while in the AWS code they represent characteristics and tolerance, respectively. These two codes may be distinguished by the fact that the first dot in the top row of the AWS code is always black. Examples of how this works out for some commonly used sizes of capacitors are given in the table that follows.

First Dot		Second Dot		Third Dot		pf	μf
Brown	(1)	Black	(0)	Black	(*no zero*)	10	0.00001
Green	(5)	Black	(0)	Black	(*no zero*)	50	0.00005
Brown	(1)	Black	(0)	Brown	(0)	100	0.0001
Red	(2)	Green	(5)	Brown	(0)	250	0.00025
Green	(5)	Black	(0)	Brown	(0)	500	0.0005
Brown	(1)	Black	(0)	Red	(00)	1,000	0.001
Orange	(3)	Black	(0)	Red	(00)	3,000	0.003
Brown	(1)	Black	(0)	Orange	(000)	10,000	0.01

The tolerance rating corresponds to the color code (i.e., red means 2 percent, green 5 percent, etc.). The voltage rating corresponds to the code number multiplied by 100 (i.e., orange means 300-volt rating, blue 600-volt rating, etc.).

Ceramic capacitors were developed as a substitute for mica capacitors and have the same characteristics—low losses and relatively high working voltages. Ceramic capacitors are made in the form of tiny disks or small cylinders. The capacitance and the voltage rating are normally printed on the body.

Paper capacitors, made by rolling sandwiches of metal foil and waxed paper into a cylinder, have capacities varying from about 0.0005 to 15 μf. Paper capacitors are probably the most widely used in radio. They are inexpensive and have relatively low leakage losses, and their range of capacitances is suited for many applications. Working voltage and values are usually printed on the body. Paper capacitors should be kept away from excessive heat. The wire lead at the end with the colored band is connected to the outside foil. If one side of a paper capacitor is to be connected to ground, this outside foil lead should be used.

A number of plastic materials have been developed as substitutes for paper for use in fixed capacitors. They offer the advantage of thinness combined with low leakage, and they permit large capacitance to be built into small space.

Capacitors of the rolled type are enclosed in simple cardboard tubes or in sealed plastic cases. Just as RF chokes have unwanted capacitance, the roll-type paper capacitor has unwanted inductance; the coiled metal foil sheets act just like turns of wire. The effect is minimized and often eliminated if the external connections to the alternate foils are made to the edges of the latter rather than to their ends.

Electrolytic capacitors, using a special chemical paste between two conducting surfaces, have high values of capacitance, yet are relatively compact. They are used in the filter circuits of power supplies where large capacitance is essential in smoothing out the ripples in rectified alternating current.

Because of their chemical makeup electrolytics possess the

property of self-healing if the dielectric film is broken down momentarily by excessive voltage. Mica, ceramic, and paper capacitors must be discarded once they have been punctured. Electrolytics must never be located too near a source of intense heat, since the film will be weakened and the voltage limit thereby reduced.

The *variable capacitor* usually consists of two sets of metal plates, one set fixed and the other movable. The movable plates are attached to a shaft so that they may be rotated between the fixed plates. The plates may be made semicircular or odd shaped, depending on the purpose for which the device is to be used.

The greater the number of plates and their area and the closer their spacing, the higher the capacitance of a variable capacitor. For receiving purposes the spacing can be small, since the voltages in receivers are quite low. For transmitting purposes the spacing is often considerable. Variables run in size from about 5 to 467 pf. Double, triple, and quadruple units are common.

The smallest variable capacitor is the *trimmer capacitor*, consisting of two conducting surfaces separated by a piece of mica. A screw forcing the plates closer together or permitting them to spring apart is used to vary the capacitance. A trimmer capacitor might vary from 1 or 2 to 5 pf. Trimmers are often used in parallel with tuning capacitors for fine adjustment.

Hardware

Large numbers of sockets, switches, plugs, jacks, terminal lugs, etc., are used in the construction of electronic equipment. These items can be lumped under the general heading *hardware*. An experimenter becomes familiar with them quickly when he undertakes the assembly of his first kit project. Three representative collections are shown here in Figs. 7-12, 7-13, and 7-14.

7-12. Tube sockets: (A) top and bottom views of 7-prong miniature, (B) top and bottom of 9-prong miniature, (C) top and bottom of "octal," (D) bottom of 4-prong, and (E) bottom of 5-prong.

7-13. A few of the switches found in electronic equipment: (A) power type, (B) shaft-operated rotary or push-pull, (C) two types of multi-position rotary, (D) lever type, (E) three-position, center "off," toggle, (F) bat-handle toggle, (G) short-handle toggle, (H) two sizes of slide type, and (I) very light action microswitch.

Used Parts

In many residential communities, a common Monday morning sight is a discarded television set sitting sadly on the curb, waiting to be picked up by the garbage collectors. If retrieved, dusted off with a brush or a vacuum cleaner, and studied

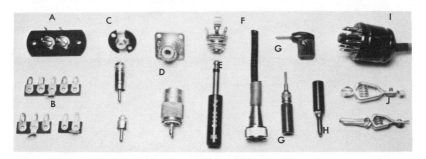

7-14. An assortment of radio hardware: (A) screw terminal strip, (B) varieties of terminal lugs, (C) phono jack with two types of phono plugs, (D) male and female coaxial fittings, (E) phone plug and jack, (F) microphone connector, (G) two styles of phone tips, (H) banana plug, (I) multiprong plug, to fit octal or other tube socket, and (J) two styles of spring clips.

carefully, this equipment can yield usable parts for replacement or experimental purposes. See Figs. 7-15 and 7-16.

If any tubes are still in their sockets, they should be wiped carefully with a dry rag and retained only if their type numbers are visible. They might or might not still have some life in them.

Components that are riveted to the chassis, such as tube sockets, terminal lugs, metal-cased capacitors, etc., are not worth the trouble of drilling out the fasteners. Attention should be paid instead to resistors, capacitors, and transformers with leads more than an inch or so long that can be snipped off readily with side-cutting pliers. Paper capacitors are worth keeping only if their marked values are still decipherable. All electrolytic capacitors are suspect and should be ignored. All resistors are desirable because they can be measured easily, as previously mentioned.

If the available set appears to be of the AC type and if the leads of its power transformer are color-coded, the transformer should be saved. The usual coding is as follows: two black wires, primary; two red wires, ends of high-voltage secondary; single red-yellow wire, center tap of this secondary; two yellow wires, rectifier filament secondary; any other pairs of

7-15. Junk? Maybe to the previous owner, but a source of small treasure to an inquisitive ham who wants parts for experimentation.

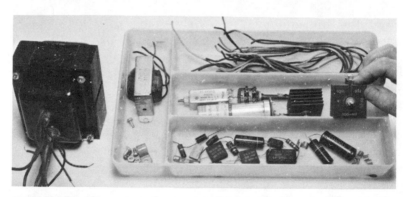

7-16. High-voltage transformer, audio transformer, rectifiers, resistors, capacitors, volume controls, switches, wiring harness—all retrieved from the chassis shown at the top of the page. Not shown is a collection of useful nuts, bolts, washers, etc.

green, brown, or slate wires, low-voltage filament secondaries.

A smaller transformer, connected between the audio amplifier and the loudspeaker, can be identified with certainty by its four leads: blue and red for the primary, green and black for the secondary.

8

Construction Techniques

The quickest and most satisfactory way to familiarize yourself with electronic components and construction techniques is to make a relatively simple piece of equipment with your own hands. Set up this project as part of your home-training program leading toward your ham license: a little code practice every day, plus a little reading of theory and FCC rules and regulations alternated with a little work on the equipment. The latter should be something that will continue to be useful after you finish it.

Without a doubt, the best bet for a beginner is a *VOM*, short for *volt-ohm-meter*. This has long been recognized as the most versatile of all electronic instruments, not only for ham purposes but also for general testing of home appliances ranging from lamps to radio and television sets and the electrical systems of automobiles and other vehicles. Two excellent meters that you can assemble from kits are described in Chapter 9.

Many helpful hints and suggestions to facilitate the work are given on the pages that follow here.

There is another way of "rolling your own." You can buy sheet aluminum or preformed blank aluminum chassis and dozens of individual components and bits of hardware, drill lots of holes in the chassis for the parts, and then assemble and wire the latter in accordance with published plans. There's nothing wrong with this method, which is widely used by experimenters who have the shop facilities and the time to do the mechanical work. However, it can be very tedious, and it does not add much to a person's education. Nowadays most amateurs start with kits, which offer the following marked advantages:

1. They use accurately preformed and punched chassis, invariably of strong steel. Bending, drilling, and filing are all eliminated, and the builder can concentrate on the more important tasks of assembly and wiring.
2. They are sure to work if assembled properly, the basic circuits having been worked out by the manufacturer's engineers.
3. As complete packages, they are cheaper than the aggregate components bought individually.
4. The instructions include both picture and schematic diagrams that any beginner can follow.
5. The completed units present a professional appearance.

Don't get the impression that kits are child's play. If you build a fifteen-tube receiver or a ten-tube transmitter, the *technical* work is the same whether you start with your own parts or an organized kit. However, the kit is much less of a gamble, yet just as challenging and instructive. Early kits were mostly for very simple items, but the whole idea of kit construction has become so popular that manufacturers now offer for amateur assembly a variety of highly sophisticated projects that cost as much as $900 and take a winter of spare time to put together.

The initial experience provided by a kit stands the builder in good stead when he decides at some later time to make changes in it or to construct special equipment of his own design (Fig. 8-1).

Tools for Radio Construction

Because the heavy job of chassis fabrication is done in advance in kits, only a few relatively small tools are needed for the rest of the work. The accompanying illustrations show these in separate groups for ease of identification. Many of them are standard hardware-store items; others are electronic specialties, more readily obtainable from radio supply houses.

8-1. The basement workbench that is almost a standard fixture in American homes is ideally suited for ham radio and other electronic construction projects. Note wall-mounted tools, within easy reach.

Fortunately, good tools are inexpensive, and last indefinitely if used properly.

In Fig. 8-2, the various pliers are easily recognized because they are common tools. *Side cutters* should be used only for snipping soft copper wire, up to and including No. 18. For heavier wire, such as No. 14 and No. 12 power lines and antenna wire, use *electrician's* pliers; in addition to thick cutting edges, these have a stong, flat nose that is useful for twisting wires and pieces of metal and for holding purposes in general. *Long-nose* pliers are intended only for light jobs, such as bending wires and positioning small parts; don't use this tool for heavy twisting or bending. *Slip-joint* pliers, sometimes called *plumber's* pliers, have concave jaws that take a good grip on cylindrical objects, such as fuse mounts and coaxial cable fittings.

It may seem odd to find a pair of tweezers in Fig. 8-2, but after you've used this tool a few times you'll agree that it's indispensable for picking up and placing thin washers, small screws and nuts, transistors, etc. Another big help, literally a "third hand," is the "seizer," in effect a very thin pair of long-nose pliers with latching handles. This clamps firmly on wires and small objects, gets them into tight spots, holds them for soldering, etc. The long-handled, spring-loaded clip at the extreme right of Fig. 8-2 does a similar job.

The nut starter is a big help for people with two or three thumbs on one hand. It is intended only for getting nuts over the first few threads of machine screws, not for tightening them down. The starter is made of soft plastic, with openings in the ends $3/16$ and $1/4$ inch in diameter. You simply squash an end over a nut, which is held by friction, and you can then apply the nut over the screw without dropping it.

No one ever has enough screwdrivers, but the collection in Fig. 8-3 makes a good start. The standard types vary in blade width from $5/64$ to $5/16$ inch, and take care of all the screws found in radio equipment. Fully 90 percent of these are No. 6 machine screws; most of the balance are the smaller No. 4 size, and the rest, usually found on power transformers, are No. 8 or No. 10. Occasionally we find some Phillips head screws, so it pays to have at least one X-point screwdriver for them.

You can be old-fashioned and try to fasten nuts with a pair of pliers, but nutdrivers do a faster and better job. The majority of nuts for No. 6 screws are $1/4$-inch hexagons; for the No. 4, $3/16$ inch, for No. 8 $5/16$ inch; and for No. 10, $3/8$ inch. The hex nuts found on volume controls, variable capacitors, toggle and rotary switches, etc., are $1/2$ inch across. A driver for this size is particularly useful because it does a quick job without leaving the slightest mark on a panel, as pliers and wrenches have a nasty habit of doing. Individual nutdrivers are cheap and are much handier than the type that uses a single handle with an assortment of detachable shafts.

Typical uses for the tools are illustrated on the pages that follow.

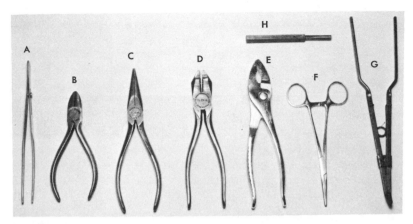

8-2. Tools for radio construction: (A) tweezers, (B) side cutters, (C) long-nose pliers, (D) electrician's pliers, (E) slip-joint pliers, (F) "seizer," latching handle gripper, (G) spring-loaded clip, and (H) nut starter.

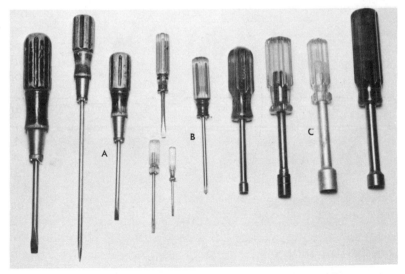

8-3. Tools for radio construction: (A) six sizes of screwdrivers, (B) Phillips-head screwdriver, and (C) four sizes of hex-nut drivers.

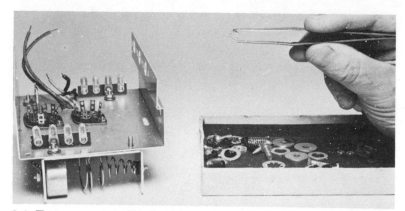

8-4. Tweezers are indispensable for lock washers, which are difficult to pick up and hold with the fingers alone. The chassis of VTVM (left) has two tube sockets and two terminal lugs already secured with small screws and nuts. One screw of the battery holder is to get the lock washer and nut and a terminal lug.

8-5. A single-lug terminal is easily placed over a screw with fingers, but a tiny lock washer goes on more readily with the aid of tweezers.

Soldering Irons

The single most important tool in electronic construction is without question the soldering iron. "Iron" is not really the right term, since all soldering tips are made of copper, but

8-6. A plastic nut starter is simply pressed over a nut to pick it up—

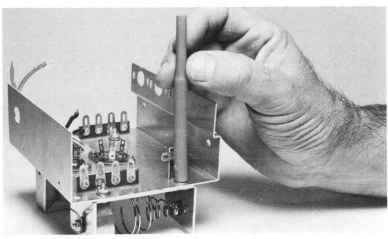

8-7. —and given a twist or two over the end of the screw to start the nut evenly on the thread. Tighten with a regular nutdriver.

that's how the tool is universally known. Two irons, as a minimum, are needed. See Fig. 8-8. The soldering *pencil* is an insulated handle with a candelabra-size socket in one end. Into this can be screwed individual tips. For most work the $5/16$-inch, 25-watt size is just right, and this is supplemented by

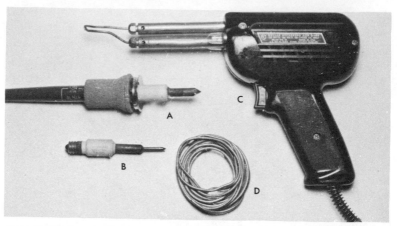

8-8. Soldering tools: (A) pencil-type iron, (B) small tip for pencil iron, (C) gun-type iron, and (D) roll of rosin-core solder.

a ⅛-inch point for very close work. These tips contain coils of resistance wire which become hot with the passage of house current through them; the heat transfers to the copper ends by conduction. Irons of this type take several minutes to warm up to a temperature that will melt solder.

The soldering gun works on an entirely different principle. This tool is actually a step-down transformer. The primary consists of several hundred turns of fine wire, connected to the power line through a trigger switch. The secondary is a single turn of brass tubing about ½ inch in diameter, the ends of which stick out from the body like the barrels of an over-under shotgun. Clamped into the free tips of these tubes is a loop of heavy copper wire. Because the primary has many turns and the secondary only one, the voltage induced in the latter is very low but the current is very high, high enough, in fact, to heat the tip to solder temperature in three or four seconds. A medium-size gun like the one illustrated is rated at 135 watts.

Aside from its quick heating feature, a gun has the advantage that its wire tip can be bent or twisted readily to make it fit into confined spaces without damaging nearby components.

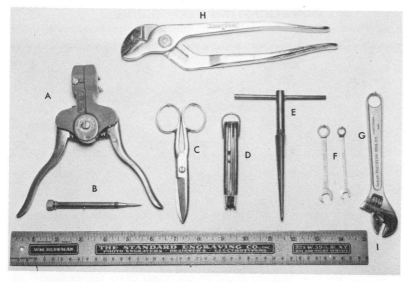

8-9. Miscellaneous tools: (A) wire stripper, (B) awl, (C) scissors, (D) knife, (E) tapered reamer, (F) "ignition" wrenches, (G) adjustable jaw wrench, (H) arc joint wrench, and (I) ruler.

The solder used in all radio work is a soft alloy of lead and tin, usually half and half, with a core filled with rosin. This is called *flux*. Its purpose is to absorb oxidation products that form on metal joints when the hot iron is applied; without it, molten solder simply rolls away without sticking.

The entire secret of successful soldering lies in two words: *cleanliness* and *heat*. If either the iron or the joint is dirty, no amount of heat or flux can make the solder adhere.

To facilitate soldering, practically all connection lugs on components and certainly all hookup wires are pretinned in manufacture. This *tinning* is a thin coating of solder, to which other solder adheres perfectly under the influence of heat.

When cold and clean, the copper tip of either the resistance or gun type iron shows its well-known characteristic color. As it warms up, however, it starts to blacken as the metal combines chemically with the oxygen in the air. The trick in tinning an iron in preparation for soldering is to turn on the juice with one hand and to hold rosin-core solder against the tip

8-10. A nutdriver (left) and screwdriver are used together to tighten a nut and bolt quickly. To keep the terminal lug from twisting, hold the nut stationary with the driver and tighten the screw.

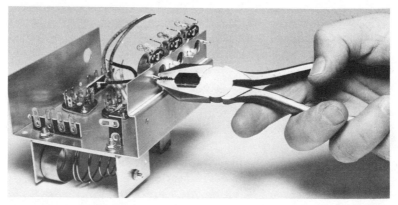

8-11. Three small potentiometers are mounted by twist lugs on their cases. For this job use heavy electrician's pliers, never the long-nose.

with the other. As the temperature rises, first the rosin melts out from the core, and then the solder melts too. Because the flux absorbs the oxidation products, the solder sticks to the tip.

Let's assume that a tinned wire has been twisted into the

hole of a tinned lug. Hold the tip of the iron and a length of solder to the joint, so that the solder runs into the latter. Be stingy with the solder; for radio purposes a thin layer is as good as a thick one. Remove the length of solder, but keep the iron in place for a slow count of five. Now pull it away, but do

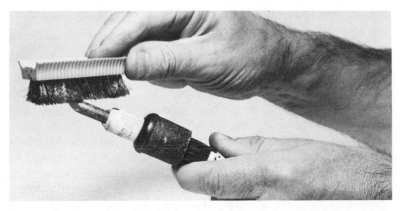

8-12. To keep the tip of the iron free of corrosion, brush it frequently with a brass brush, and apply a dab of solder if necessary to renew

8-13. The tip must be smooth and free of pits. Occasional dressing with a fine file is necessary.

not disturb the joint: the solder takes a few seconds to set solid.

Heat is needed to cook out excess flux, in the form of bluish smoke. If the iron is removed too soon, the flux can readily form an insulating film between the very surfaces the solder is supposed to join. This is called a *cold joint*, for obvious reasons. Under common circumstances it can present either a measurable high resistance, or, astonishingly, an absolutely open circuit!

Beginners have a tendency to pull away the iron the instant the solder starts to melt, thinking that the heat will damage the component. Radio parts in general are built to take it; too little heat is worse than too much. Transistors are an exception in this respect, and some precautions about handling them are included in Chapter 4.

An important fact to keep in mind in all soldering work is that molten solder is as fluid as water and therefore obeys the law of gravity. Wherever possible, turn or prop up a chassis or other assembly so that the solder runs *away* from the joint. See Fig. 8-15. This is especially necessary with multiposition rotary switches. Their contacts and terminals are often quite close, and it doesn't take much excess solder to bridge them.

If a resistance-type iron is left on but idle for any length of time, while connections are being prepared, the continuing heat is likely to burn the tinning. If this isn't too far gone, a few quick strokes with a brass brush (sold for cleaning suede shoes) and a dab of solder might restore the shiny color. If this doesn't do the trick, clean the tip with a fine file and apply solder immediately (Fig. 8-12). With extended use, a tip becomes pitted. Let it cool, file it down, plug it in again, and retin. See Fig. 8-13. Eventually, of course, the copper just burns away.

The wire tip of the soldering gun is much less subject to corrosion because its "on" cycle is very short and most of the time the current is off altogether. However, it does need occasional cleaning and shaping.

Soldering tips have an annoying habit of expiring at 10:15 P.M., when all local stores are closed and a project is only a

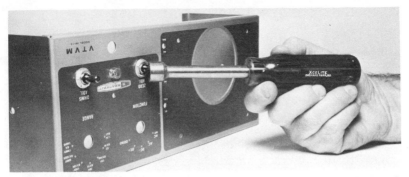

8-14. A nutdriver in ½-inch size is the only tool that should be used on hex nuts of potentiometers, switches, etc. It not only gives good leverage but it also leaves the panel unmarked.

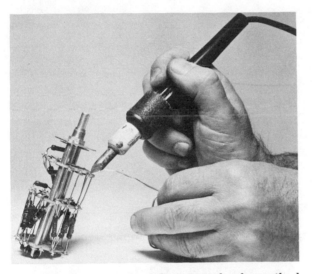

8-15. The range switch of the VTVM has been tilted slightly so that molten solder runs away from the lugs, instead of into the close-spaced contacts.

dozen or so connections away from completion. They are cheap, and at least one spare of each type should be kept on hand for just such emergencies.

Stripping Insulation

In radio construction, countless wires must be cut to length and the insulation removed from their ends. The conventional method of stripping is to use a knife and to pare off the covering in the manner of sharpening a pencil. This can be difficult and frustrating, because the soft copper wires, especially those in stranded conductors, are too easily nicked or severed by a sharp blade. A much smarter idea is to use a *wire stripper*. This magical tool works like a pair of cutting pliers with a reverse action. See Figs. 8-16 and 8-17. With the handles open,

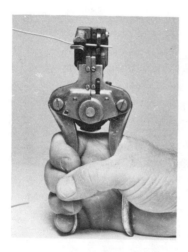 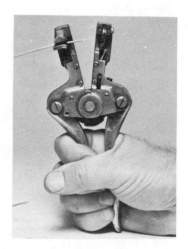

8-16. A wire stripper is a great time-saver. In this first step, the insulated wire is in the closed jaws, but the handles are open.

8-17. When the handles are squeezed, the insulation is cut, the jaws open, and the insulation falls away.

you put the wire horizontally through two sets of jaws. One pair grips it; the other has hardened blades that will cut only the insulation. Squeeze the handles. First the blades go through the insulation, then the other jaw pulls the wire away to the left, causing the severed insulation to slide off and drop away. Four standard sizes of hookup wire can be accommodated in the jaws.

Component Support

Practically all small resistors and capacitors are supported by their own leads. These are merely cut to length and soldered to various lugs on sockets, terminal strips, etc. See Fig. 8-18. In most cases the leads are left bare, as they are fairly stiff and do not change position once they are secured. If several components are in a tight cluster, it is good practice to cover the wires with an insulating fabric tubing long known as *spaghetti* because it looks just like that delectable pasta.

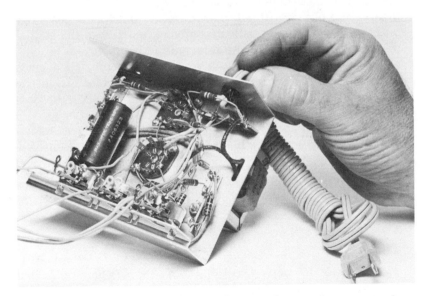

8-18. A typical small radio chassis, showing how resistors, capacitors, and other small parts are mounted by their own leads.

9

Test Meters You Can Make

As mentioned in the preceding chapter, the most useful test instrument for ham radio and other hobby purposes is one that measures volts, ohms, and amperes. In its simplest and most widely used form it is called a *VOM*. This is a contraction of *volt-ohm-meter* and is pronounced *vee-oh-em*. Technically, the term *multimeter* also applies, but this is generally reserved for more sophisticated meters, usually of the digital display type. Both types are interesting kit projects at money-saving prices. For example, a representative VOM kit costs about $75 in kit form and $94 factory assembled, from the same manufacturer; for a digital model, the kit is $70, the factory job $100.

If you are just starting in ham radio you will be very happy with a VOM. If you have had some experience with other electronic gear such as tape players, tuners, amplifiers, etc., you will probably be attracted to the digital meter. A typical instrument of each type is described on the following pages.

The VOM

The heart of the VOM is a highly sensitive DC ammeter that can measure incredibly small currents down to a few microamperes; that is, *millionths* of an ampere. The top scale reading for this range is 50 microamps. Other DC measurements are made with resistors of various values in *shunt* with

9-1. The Heathkit Model IM-105 VOM includes all necessary parts and hardware down to lock washers. The meter itself is part of the front panel. The "chassis" consists of the two square printed circuit boards, left and right of the front panel.

9-2. Back side of a circuit board. Soldered leads of components, inserted from front, are being snipped off closely.

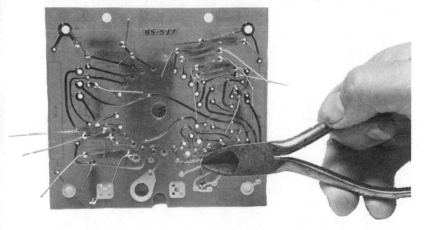

the meter (that is, connected directly across its terminals) for higher current values, and with additional resistors in *series* with the meter for all voltages. There is no provision for measuring AC amperes, as there is little or no need for this in radio work. However, AC voltage capability is made possible by the use of small diode rectifiers that change the AC to be

9-3 and 9-4. Above: Both circuit boards, completely assembled and wired. The elliptical objects in the center are rotary switch sections, which will be controlled by a single shaft through the center holes. Note how small components fit snugly against board surfaces. Left: The circuit boards are "mated" by stand-off collars in their corners. This assembly fits directly behind front panel of meter. Remaining wires are connected later.

measured into DC. All functions are selected by a rotary front-panel switch that has twenty positions.

For DC volts, the scales read 0-.25, 2.5, 10, 50, 250, and 500 full scale; for AC volts, the same except no .25 range. For DC amperes, in addition to the 0-50 microamp range mentioned previously the scales read 0-1, 10, 100, and 500 milliamperes (thousandths of an ampere) full scale, and one 0-10 ampere range. DC and AC voltage readings up to 1,000 and 5,000 volts are also available for checking some of the circuits of high-

9-5. Back view of completed meter, minus cover plate. Two small batteries at top are needed for measurement of resistance.

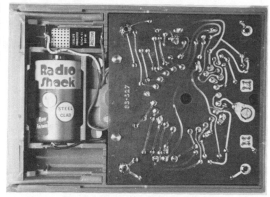

power transmitters. This is rather risky business, and should be avoided.

For resistance measurement two small batteries inside the case are switched to the basic meter through a network of resistors, with one variable. When the probes of the meter's flexible test leads are touched together (that is, short-circuited), the variable is adjusted so that the meter needle reads zero. When the leads are separated (that is, open-circuited), the needle swings to the extreme other side of the scale for a reading of infinity, which is no reading at all.

The top scale of the meter is calibrated in ohms and reads from 0 to 2,000 directly when the selector switch is in the X1 setting. Four other positions are marked X10, X100, X1K, and X10K. For these positions you must multiply the scale reading by 10, 100, 1,000, or 10,000, respectively.

For testing an unknown resistor, start on the X1 range and connect the test probes to the leads. If the needle does not move, the value is more than 2,000 ohms, so switch to the X10 range. If there's still no action, try the next range. If there is no response on any of the scales, the resistor is merely burned out or broken.

In many cases the actual resistance of a resistor, a length of wire, a printed circuit-board line, etc., is not important. What you need to know is whether the circuit through it is complete or open. This is called *continuity checking.*

When measuring voltages always start with the selector

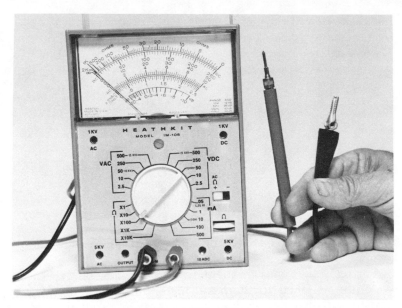

9-6 and 9-7. Above: Completed VOM. Resistance setting is at ×100. With test leads apart, pointer is at rest. Below: Connected leads give effect of short-circuit, so pointer correctly reads zero ohms.

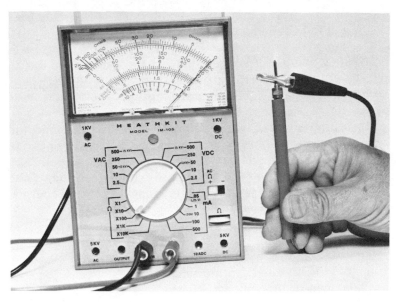

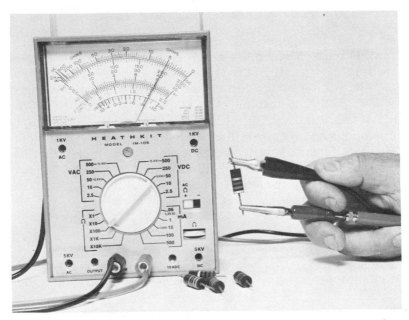

9-8 and 9-9. Above: Resistor reads 5 ohms. With ×100 setting, this means value is 500 ohms. Below: Is battery up to snuff? Yes. It reads 9 volts on 10-volt DC scale of meter.

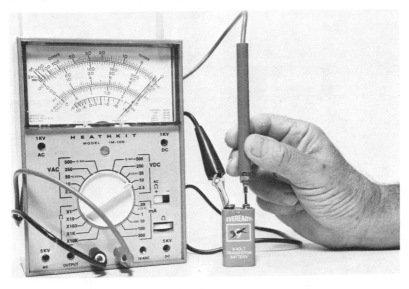

switch at the highest setting. If there's no immediate indication, you can move down a setting at a time until you get a reading. If you start at a low setting and unexpectedly encounter a high voltage, the meter needle might be damaged when it bangs against one of its stops.

Assembly and wiring of the VOM from its kit are simple operations and require ordinary hand tools and a clean soldering iron. Open all little boxes and envelopes, identify all the components and pieces of hardware, and allow yourself several evenings for the job.

Solid-State Digital Meter

The first time you inspect almost any piece of ham equipment containing transistors you will realize quickly that both they

9-10. Erratic resistance reading on control switch of mike indicates dirty internal contacts. In good condition, switch should read open when off, zero ohms when pressed down. ·

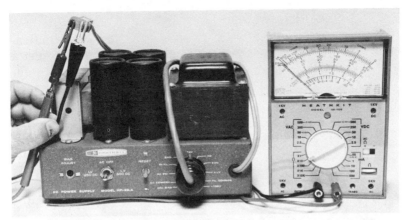

9-11. AC power pack of transmitter dead? Maybe it's only a blown primary fuse or open circuit breaker. Continuity check on line cord, with transmitter control switch "on," will be good clue.

and most of their other circuit components are very small and are usually mounted close together on printed circuit boards. The type numbers of the transistors are barely visible and the thin connection leads are not marked at all. To identify the latter you must look for an indexing dot or a flat area on the rim of the body. About 98 percent of all resistors and capacitors do not bear numbers of any kind, but must be identified by their color markings. If you don't see well close up or are slightly color-blind you should take up another hobby!

An inexpensive addition to your tool kit will help you with this problem; it is nothing more than a common "reading glass," about three inches in diameter. A stamp magnifier also serves the purpose. So armed—or rather "eyed"—you are ready to tackle a transistor construction job as part of your technical education. It should be something of real value; not a mere exercise. A project that certainly qualifies is the Heathkit Model IM-2202 Digital Multimeter, shown here in various stages as it was assembled by the editor of this book.

After looking at the picture of the completed meter, Fig. 9-13, you will probably ask, "Where is the meter of this multimeter?"

9-12. Good light and a magnifying glass are necessities for anyone working on solid-state equipment. The item under scrutiny is a printed circuit board of the meter shown on the next page.

There isn't any! Depending on how you punch four buttons and turn an eleven-position rotary switch on the front panel, all readings of volts, amperes, and ohms appear magically as bright red numbers in the rectangular space above the buttons; the decimal point is in the right place and a + or − sign shows the polarity of the connecting leads to an object under a voltage test. The instrument gets its name from the word *digit*, which means any number from 0 to 9. The circuitry that accomplishes this magic is very complex but reliable.

As with most kit projects, the assembly and wiring of this meter requires only a few simple hand tools. It also requires a lot of patience, because the transistors and other parts are not only very small but also very numerous, and usually have to be handled with tweezers rather than bare fingers. This is all part of the game. While the instruction book that comes with the kit is well prepared, you'll probably take five minutes to puzzle out the hairlike leads of the first transistor you pick up. Is that tiny dot on it the index mark? Is the lead nearest to it the B (base)? Is the E (emitter) wire on the left or right? You have to decide; the third lead then automatically becomes the C (collector).

Because they have only two leads, diodes appear to be easier to install. However, you need good light and a clean magnifying glass to determine which end of the body bears an identifying band. The purpose of all diodes is to allow current to pass in one direction and to block it in the other. If you connect one backwards the circuit goes completely out of kilter.

You shouldn't attempt any assembly work the first night. Be satisfied to unwrap all packages and to check their contents against the parts lists in the instruction book. This is quite a job by itself, since there are about 225 solid-state devices, resistors, and capacitors, plus scores of other bits and pieces that can be lumped as hardware.

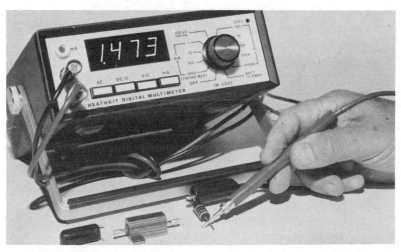

9-13. Completely solid-state, this Heathkit Model IM-2202 digital-type test meter features direct readout on all its measurement functions.

9-14 and 9-15. Above is a topside view of two of the circuit boards of the IM-2202 digital multimeter, as furnished with the kit. The leads of all components are pushed through holes and soldered on the underside. Below are the same boards after five evenings of assembly and soldering! The holes in the lower area of the board on the left will later take a six-section function switch, controlled by a knob on the front panel.

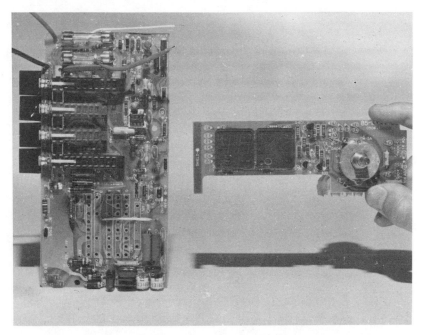

9-16 and 9-17. Above: At the right is a small board that becomes part of the front panel of the meter. The two dark rectangles at the left are the readout displays. Below: Small board is now mounted to main printed circuit board, but the wiring is not yet complete. Function switch is at left.

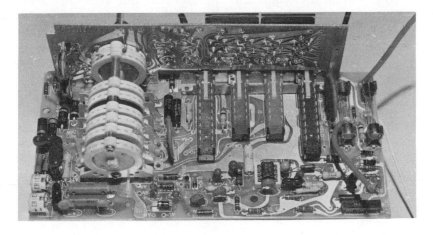

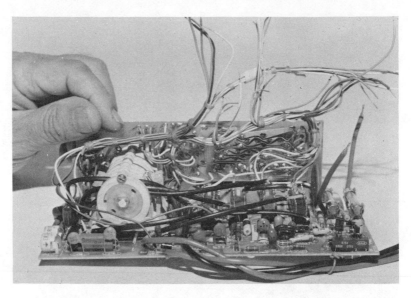

9-18. Connections between this assembly and the other circuit boards of the meter are made with the loose harness of wires.

The Printed Circuit Board

In modern solid-state equipment only about 10 percent of the connections are made with ordinary wire that you must cut to size. The other 90 percent are made by metallic lines printed on one side of a thin board of insulating material. On the other side, the lead wires or soldering lugs of most of the components are pushed through holes and then soldered to the printed lines. This saves a large amount of work, but, because of the closeness of the lines and the danger of solder spill-over, it forces you to develop a very delicate soldering technique, which is all to your advantage. Adeptness comes only with practice. Of course, you will keep the magnifier close by and use it religiously after doing a group of four or five joints.

This multimeter has its major components mounted on three flat circuit boards and the remainder on an L-shaped bottom section of the carrying case. The smallest board mounts in

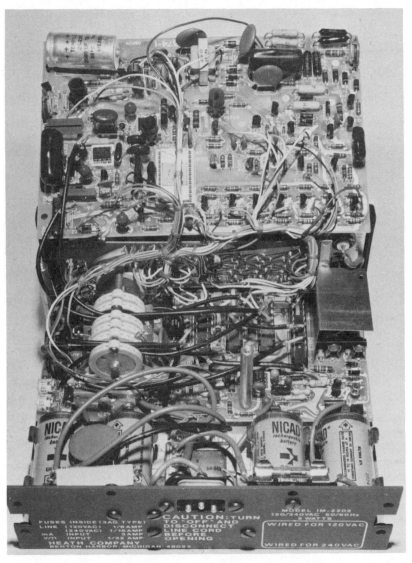

9-19. Back view of the wired multimeter. Obviously there is no wasted space here! The topmost circuit board is folded down and is secured parallel to the base of the carrying case by screws into two upright posts.

a vertical position directly behind the front panel of the case and the other two are arranged as an open horizontal sandwich over the case's bottom. The fit is very tight, but fortunately the meter can be turned on and checked for operation before the top half of the sandwich is screwed down.

An interesting feature of the meter is its self-contained combination of four nickel-cadmium "D" size storage cells and a charging circuit. The batteries provide up to eight hours of field operation before they need recharging, and the latter is accomplished by plugging the instrument into an AC outlet. For bench use, merely leave it plugged in.

All measurement functions are protected against overloading by means of fuses or diodes. The carrying handle also serves as a stand for the meter, angling the front panel upward for easy observation of the readouts.

10

The Code Is a Must

A basic requirement for an amateur radio license is the ability to send and receive simple words and sentences as represented by the dots and dashes of the *International Morse Code*. It is incorrect to call this the *Morse Code*, because the latter is quite different in make-up. It was the system used for almost a hundred years on the *wire*-line telegraph circuits of the United States and Canada, but in Europe it was replaced in the early 1900's by a much simpler arrangement of the dot-dash characters called the *Continental Code*; this was carried over to *wireless* telegraphy as Marconi advanced the state of the art. The word Continental gave way to International when radio became a worldwide means of communication, but the original term survives. Many operators still refer to the code simply as Continental.

People taking up ham radio generally want to talk into a microphone, and they write peevish letters to their congressmen complaining about the FCC when they learn that the code is an absolute must for a license. They struggle through the dots and dashes, pass the written exam, get their licenses, and descend on the ham stores to buy equipment. They go on the air, and then many of them undergo a strange change of altitude. They discover that Continental is not merely a code but almost a new language, that it is far more effective than voice for DX, that it is the mark of the sophisticated operator, and all-in-all that it is a lot of fun. It also solves some serious language problems. For instance: How do you communicate with a Japanese ham if he doesn't know English and you don't know Japanese? You can't exactly talk to him, but with the help of the international "Q" signals (see Chapter 16) you can exchange a lot of information on CW.

Learning the code is entirely a matter of motivation. You want to drive a car? You practice with a licensed driver next to you and you memorize the state's highway regulations. You want to run a ham rig? You practice the dots and dashes and study the rules of the FCC.

The International Morse Code

This uses only two sounds: a very short one called a *dot* or *dit* and a slightly longer one called a *dash* or *dah*. As you scan the accompanying chart, Fig. 10-1, you'll probably observe that except for the figures the combinations appear to be arbitrary. They are. Note also that the letter O and the figure 0 are different. When numbers must be transmitted accurately, they are usually spelled out completely as words. For example: one four two seven three kilohertz.

Because of the informal nature of CW operation there is not much need for punctuation marks, but you are sure to hear some of the five shown in the chart. The question mark is particularly useful with the "Q" signals, and the slant bar with call signs. For example, a second district licensee like W2PF, operating in Forida, which is part of the fourth district, would identify himself as W2PF/4.

Three of the letters have additional meanings. R is the common signal of receipt, a natural choice. K means *go ahead*, an invitation to transmit. Z, a carry-over from military procedure, is often a suffix to replace the letters GMT or UTC after a four-digit readout in the 24-hour clock system; for example, 0745Z, 1653Z. (See Chapter 17.)

A few foreign characters are also included because you are quite likely to run across them when you tune for CW stations for code practice. The German dah-dah-dah-dah is an especially handy short form for the common combination CH.

International Morse Code

Punctuation

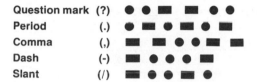

Question mark	(?)	
Period	(.)	
Comma	(,)	
Dash	(-)	
Slant	(/)	

Figures

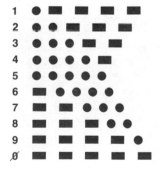

Foreign Characters

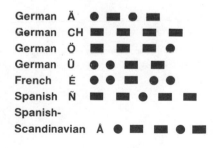

German	Ä
German	CH
German	Ö
German	Ü
French	É
Spanish	Ñ
Spanish-Scandinavian	Å

Fig. 10-1

10-2. This Heathkit code oscillator contains few parts, and is a simple assembly and soldering job. Small round object, in hand, is the loudspeaker. At lower right is "straight"-type telegraph key, perfectly suitable for live CW transmission on the air with standard amateur type transmitters/transceivers. (See Chapters 12 and 16.)

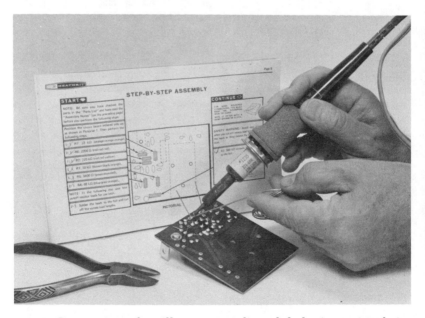

10-3. Components of oscillator mount through holes in a printed circuit board. "Pig tail" leads are soldered to metallic printed lines and then cut off close to the surface. Low-wattage iron with slender tip is needed for this work. Board is tilted so that molten solder runs away from, not across, adjacent lines.

Practice Does It

There are no tricks or shortcuts to learning the code, any more than there are to learning to swim, play tennis, drive a car, or fly a plane. It is a matter of planned practice and personal determination.

What you need right away is a code oscillator of some kind with which you can make your own dits and dahs. This should have an earphone jack, so that you can practice to yourself with phones on, without disturbing others in the house. The oscillator shown in Figs. 10-2, 10-3, and 10-4 is a Heathkit you can put together in an hour; there are several others like it on the market. It uses transistors, is powered by a single 9-volt battery, comes with a perfectly good telegraph key that you can use later with a transmitter, and emits a pleasant tone.

The proper position of the hand on the key is shown in Fig. 10-5. Rest your whole arm on the table, and relax. The key takes very little pressure, and should be pressed rather than tapped.

Adjust the spacing of the key's contact points by means of the knurled screw at the back end of the lever, and the spring pressure by the screw nearest the knob. Rest your first two fingers on the latter and work it with a light tapping motion. At first you'll probably slap it too hard and run dots and dashes together, but you'll master the technique quickly. Keep the code chart in front of you and start with the easiest characters, which are those only with dits: E, I, S, H, 5. You should include the figures right away, as all ham call signs contain one.

Make up some simple words or combinations such as HIS, SIS, HE, SHE, SEE, 5S, etc. These are pretty easy, so you might proceed immediately to the dash characters T, M, O, Ø. The dash should be noticeably longer than the dot, but don't drag it out. Now make up more words, first with dashes alone and then with combinations: TO, TOO, TOM, THE, THIS, SIT, SEE, HOT, HOSE, THEM, etc.

10-4. Loudspeaker is mounted on rear edge of chassis, and projects sound through open back of cabinet. Small battery, of type found in transistor receivers, fits behind speaker. As it is used only intermittently, it can last for a couple of years.

The next day repeat the routine. Scramble the words several times. Send them first with your eyes on the chart, then again with your gaze averted.

You will understand the importance of uniform spacing between the dots and dashes of letters when you progress to the rest of the alphabet. For instance, if you falter after the dash of the letter B what you actually send is TS. If you inadvertently separate the dash from the dots of 6 you produce TH.

Try to maintain daily practice sessions, even if they are only fifteen or twenty minutes long. This is much better than longer sessions separated by longer periods.

Can you really learn the code when you do both the sending and receiving? Yes. Would-be hams have been doing it for decades. However, the process certainly is much shorter if two persons start it together, taking turns at the oscillator. And a group isn't limited to two. There are numerous cases of whole families, including both parents and as many as five children, all participating at the dining room table, and all qualifying for their licenses!

10-5. Key is mounted on board to keep it steady. Note student's arm full length on table, two fingers on key. Earphones are plugged into oscillator for private listening. Newspaper provides endless "copy" for practice sending.

If you own or can borrow a shortwave receiver you can put it to good use by tuning in the code-practice transmissions of W1AW, the headquarters station of the American Radio Relay League, in Newington, Connecticut. These are broadcast on a number of frequencies at various hours of the day, according to schedules published in QST, the League's monthly magazine and the bible of the amateur fraternity. The text material is taken from the pages of the magazine, so you can check it readily. The keying is done accurately at various speeds by a tape sender.

There are literally thousands of other CW stations on the air, and by tuning around you can always find one sending at a rate you can follow.

If you own a tape recorder, you may find it helpful to record the timed W1AW transmissions and to play them back at your convenience, repeating them if necessary until you are able to copy everything correctly. Figure 10-6 shows the easiest way of making the recording without disturbing the receiver setup; hang the microphone of the tape machine over the loudspeaker and let 'er run.

Commercially made tapes of practice material at various speeds are popular with people who don't want to buy receivers or transceivers until they obtain their FCC tickets. Members of ham clubs who use these tapes generally swap them until everybody has learned the contents more or less by heart, and then they dispose of them to other newcomers.

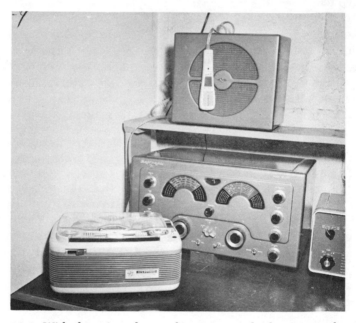

10-6. With the microphone of a tape recorder hung over the loudspeaker of a shortwave receiver, you can easily record CW transmissions from numerous stations, both amateur and commercial. Play them back at your leisure, and you'll soon respond to the rhythm of the dits and dahs.

11

The FCC
Amateur Licenses

While no license of any kind is needed for the use of short-wave receivers, the operation of shortwave transmitters is very strictly controlled by the various governments of the world in accordance with international treaties. In the United States the administering agency is the Federal Communications Commission, better known as the FCC. A license from the FCC is absolutely necessary for two-way amateur communication.

The Commission issues a valuable bulletin on the hobby which is reproduced here because it contains quick answers to the questions most often asked by newcomers.

The Amateur Radio Service, generally known as "ham" radio, is one of the oldest radio services, dating back to almost the turn of the century. It is intended for people with a technical inclination and a personal interest in the radio art. Business and commercial communications are strictly forbidden. Amateur radio is international in scope. Radio amateurs can communicate with other amateurs next door or throughout the world. Amateurs can design and build their own equipment. Amateurs communicate using International Morse Code, voice, teleprinting and television as well as experiment with other modes of communication.

The Rules and Regulations governing the Amateur Radio Service are designed to provide the service with a fundamental basis and purpose as follows:

(a) Recognition and enhancement of the value of the amateur service to the public as a voluntary non-commercial communication service, particularly with respect to providing emergency communications.

(b) Continuation and extension of the amateur's proven ability to contribute to the advancement of the radio art.

(c) Encouragement and improvement of the amateur radio service through rules which provide for advancing skills in both the communication and technical phases of the art.

(d) Expansion of the existing reservoir within the amateur radio service of trained operators, technicians, and electronics experts.

(e) Continuation and extension of the amateur's unique ability to enhance international good will.

Obtaining a License

Practically anyone, regardless of age, is eligible for an FCC amateur operator and station license. Aliens must have a valid U.S. address and not be a representative of a foreign government. To obtain an amateur license, applicants must demonstrate their qualifications in the following 10 areas: International Morse Code, rules and regulations, radio phenomena, operating procedures, emission characteristics, electrical principles, practical circuits, circuit components, antennas and transmission lines, and radio communication practices.

Operator Classes

There is a ladder of five classes of amateur radio operator licenses. You must hold or qualify for the previous class of license on the ladder to be eligible for the next higher class. There are many study aids available from stores and libraries. Amateur radio clubs, schools, and community colleges often offer amateur license preparation classes.

The first rung of the ladder is the *Novice Class*. This license authorizes beginners to acquire practical operating and technical experience in high-frequency amateur operation.

The *Technician Class* license is the second rung. It authorizes privileges on the 6- and 2-meter bands and all amateur frequencies above 220 MHz. Technicians also have the full privileges of Novice Class licensees.

The *General Class* license has privileges on all or portions of all amateur bands.

The *Advanced Class* license provides for all amateur privileges except the exclusive sub-bands of the Amateur Extra Class.

The highest and most prestigious class is the *Amateur Extra Class*. The privileges of the Amateur Extra Class license include exclusive sub-bands, and an Amateur Extra Class certificate issued upon request from the FCC Field Office where the Amateur Extra Class examination was taken.

Taking Examinations

All applicants for Technician and higher class licenses must take their examinations at an FCC Field Office or examination point. Examinations for the Novice Class license are given only by mail under the supervision of an eligible volunteer examiner selected by the applicant. A volunteer examiner must be at least 18 years of age, must not be related to the applicant, and must hold a General, Advanced or an Amateur Extra Class operator license.

All amateur operator licenses are valid for five years and are renewable. The following table outlines the ladder of amateur license classes and their examination requirements:

Class	Telegraphy requirement	Written examination
Novice (*entry level*)	Element 1A (5 wpm*)	Element 2
Technician	——	Element 3
General	Element 1B (13 wpm)	——
Advanced	——	Element 4-A
Amateur Extra	Element 1C (20 wpm)	Element 4-B

wpm means words per minute.

Applicants for amateur examinations should inquire at the closest FCC Field Office for the time and dates of amateur examinations. Amateur examinations are available to shut-ins and other persons physically unable to travel to a Commission Field Office, as shown in a physician's certification. In such cases, applicants should request the local FCC Field Office to make the necessary arrangements.

Copies of the FCC Rules for amateur radio (Part 97) are available from the Government Printing Office, Washington, D.C. 20402, for $1.50. Request Amateur Radio Rules Part 97, stock no. 004-000-0035-0. FCC Rules for the Amateur Radio Service are also available from many radio equipment stores and amateur radio publications.

Illegal Operation

A person must have an amateur radio operator license to be the control operator of an amateur radio station. Operation of a transmitter in the amateur radio service without the proper amateur license can result in penalties. Failure to make timely response to an Official Notice of Violation may also result in suspension or revocation of an amateur license and monetary forfeitures. (For the addresses and telephone numbers of the FCC field offices, see Appendix E.)

Breakdown of the Ham Bands

The differences among the five classes of ham licenses are in the operating privileges they offer and the restrictions they impose. To understand these fully, you must first familiarize yourself with the authorized types of emissions, as represented by short alpha-numeric symbols:

TYPE OF MODULATION	TYPE OF TRANSMISSION	SYMBOL
Amplitude (AM)	With no modulation.	AØ
	Telegraphy by on-off keying of the carrier wave; i.e., CW code.	A1
	Telegraphy by an audio tone-modulated carrier; usually called ICW.	A2
	Radiotelephony. Includes SSB, as this is a form of AM modulation.	A3
	Facsimile. Transmission of still pictures.	A4
	Television. Slow-scan type.	A5
Frequency (FM)	Radioteletype by shifting a steady CW carrier up or down in frequency about 850 Hz to make the "mark" and "space" impulses that operate the machines. Usually called FSK for *frequency shift keying.*	F1
	Radioteletype by modulating a steady carrier with two audio frequencies. Called AFSK.	F2
	Radio telephony.	F3
	Facsimile	F4
	Television	F5
Pulse		P

"With no modulation" usually means that the carrier is turned on briefly for control purposes.

It pays to memorize the meanings of all these symbols, because they frequently appear on the written tests for licenses.

The next table shows what kinds of transmissions may be used on what ham frequencies, but not necessarily by all licensees:

BAND	EMISSIONS	BAND	EMISSIONS
Kilohertz		*Megahertz*	
1800–2000	A1, A3	21.000–21.450	A1
(160 meters)		(15 meters)	
3500–4000	A1		
(80 meters)			
3500–3775	F1	21.000–21.250	F1
3775–3890	A5, F5	21.250–21.350	A5, F5
3775–4000	A3, F3	21.250–21.450	A3, F3
7000–7300	A1	28.000–29.700	A1
(40 meters)		(10 meters)	
7000–7150	F1	28.000–28.500	F1
		28.500–29.700	A3, F3, A5, F5
7075–7100	A3, F3	50.0–54.0	A1
		(6 meters)	
7150–7225	A5, F5	50.1–54.0	A2, A3, A4, A5, F1, F2, F3, F5
7150–7300	A3, F3	51.0–54.0	AØ
14,000–14,350	A1	144–148	A1
(20 meters)		(2 meters)	
14,000–14,200	F1		
14,200–14,275	A5, F5		
14,200–14,350	A3, F3		

On the following bands the sky is literally the limit, because virtually all types of emissions are permitted to encourage experimentation:

Megahertz: 144.1–148, 220–225, 420–450, 1215–1300, 2300–2450, 3300–3500, 5650–5925.

Gigahertz: 10.000–10.500, 24.000–24.250, 48.000–50.000, 71.000–76.000, 165.000–170.000, 240.000–250.000. *Gigahertz*, pronounced *jiga*, is a thousand million or 10^9 hertz. There aren't many hams operating in these ranges as yet, but not long ago the same thing was said about the 220–225 and 420–450 megahertz bands, which have become quite busy.

There's something interesting about the foregoing tables. Go over them slowly and note that A1 transmission (CW code) is

11-1. Surprisingly few newcomers to the ham game know that a form of real, live television called "slow scan" is enjoyed by several thousand amateur operators, about half of them in the United States and the balance elsewhere. The required camera and electronic equipment is complicated and a bit expensive, but these factors don't bother ardent experimenters. Depending to some extent on atmospheric conditions, the images can range from poor to excellent; here is a sample of the latter taken by John Smetona, K3SLJ, of Pottsville, Pa., of Allen Schulman, WØDRT, of Ballwin, Mo. During the early stages of commercial television many viewers were happy with pictures of far inferior quality!

authorized for *all* parts of *all* bands, whereas A3 (voice) is permitted only in limited parts. This is in line with the long-standing concept that the International Morse Code is the international language of radio communication.

License Privileges

Novice Class licensees may use *only* A1 in slices of four bands: 3700–3750 kilohertz, 7100–7150 kHz, 21.100–21.200 megahertz and 28.100–28.200 MHz. If you think these slices don't amount to much, you should listen on them and note how busy they are with both local and long-distance contacts.

Technicians have all authorized privileges including A3 on all frequencies above 50 MHz (this includes the popular 6-meter band), and 144–148 MHz, and they retain their Novice allotments in addition.

Up the FCC ladder, Generals get most but not quite all of the remaining bands, the exceptions being five slices reserved for Advanced and Amateur Extras: 3800–3890 kHz, 7150–7225 kHz, 14,200–14,275 kHz, 21,270–21,350 kHz, and 50–50.1 megahertz. They also retain earlier allotments.

The final slicing of the frequency pie gives the Amateur Extras exclusive rights to 3500–3525 kHz, 3775–3800 kHz, 14,000–14,205 kHz, 21,000–21,025 kHz and 21,250–21,270 kHz.

From this you can see that it pays to practice the code and study technical literature and to qualify as soon as possible for the Amateur Extra license. With this ticket you can operate anywhere in any band with all the specified forms of transmission.

Confused by some of the frequency numbers? The key is the use in them of periods or commas, or neither. FCC Part 97 of the Official Rules and Regulations, the bible of the amateur hobby, is not altogether consistent in itself, but the following examples should clarify the matter.

With numbers up to four digits there are no periods or commas: 60 hertz, 825 hertz, 1500 hertz, 600 kilohertz, 3525 kilohertz. With five digits, a comma between the second and third means that they are to be read as whole numbers without decimal fractions: 14,500 kHz means *fourteen thousand five hundred kilohertz*. This can get awkward. It is often more expedient to express these high radio frequencies as megahertz rather than kilohertz. Turn the comma into a period and 14,500 kHz becomes simply 14.5 MHz, or *fourteen point five megs*, for short. The period is used in the same sense in gigahertz figures.

The Licensing Routine

You must understand that the FCC *does not* give Novice Class examinations in any of its offices. This job is handled by volunteers who are hams themselves. The exact wording of the FCC regulations in this regard is reprinted here so that there can be no misunderstanding.

Unless otherwise prescribed by the Commission, examinations for the Novice Class license will be conducted and supervised by a volunteer examiner selected by the applicant. The volunteer examiner shall be at least 18 years of age, shall be unrelated to the applicant, and shall be the holder of an Amateur Extra, Advanced, or General Class operator license. The written part of the Novice examination, Element 2, shall be obtained, administered, and submitted in accordance with the following procedure:

Within 10 days after successfully completing the telegraphy examination Element 1(A), an applicant shall submit an application (FCC Form 610)* to the Commission's office in Gettysburg, Pennsylvania 17325. The application shall include a written request from the volunteer examiner for the examination papers for Element 2. The examiner's written request shall include (i) the names and permanent ad-

---

* Obtainable in person from an FCC office, by mail from Gettysburg, from ham supply stores, or from the examiner himself.

dresses of the examiner and the applicant, (ii) a description of the examiner's qualifications, (iii) the examiner's statement that the applicant has passed the telegraphy element 1(A) under his supervision within 10 days prior to submission of the request, and (iv) the examiner's written signature.

The volunteer examiner shall be responsible for the proper conduct and necessary supervision of the examination. Administration of the examination shall be in accordance with the instructions included with the examination papers.

The examination papers, either completed or unopened in the event the examination is not taken, shall be returned by the volunteer examiner to the Commission's office in Gettysburg, Pa., no later than 30 days after the date the papers are mailed by the Commission.

The date of mailing is normally stamped by the Commission on the outside of the examination envelope.

Of course, the question is already on your lips. "How do I find a volunteer examiner?"

Your best bet is through ham clubs, because they are always glad to have new members, and one way of getting them is to help them acquire their FCC "tickets." Check your local newspaper for meeting places and dates. Investigate schools; many have ham clubs or offer courses in amateur radio as part of their "continuing education" programs. Try National Guard and Army, Air Force, and Navy active or reserve units; they're bound to have hams among their communications personnel. Inquire at police and fire departments; many of them have hams in their dispatcher or maintenance crews.

The examiner will give you the code either with a code-practice oscillator and a hand key or with a prerecorded tape. If you flunk it, you must wait 30 days for another shot at it. If you pass, you take the written test as prescribed above. The examiner does not grade your paper as you wait; he has to send it to Gettysburg. If it clears there, your license will be sent to you directly by mail. If it doesn't pass, you'll be notified, and in 30 days you can try again.

For the higher class examinations you must appear at an FCC office or examination point. The exceptions made for the physically disabled were mentioned earlier in this chapter, in the FCC bulletin. No fees of any kind are required. Just come with a loaded fountain pen and your Novice ticket. In going from Novice to Technician you get credit for the 5 wpm code test that you took for the former, so all you have to take now is element 3, a written test. This will be graded right away, so you will know how you stand.

Suppose you pass. A very interesting possibility now presents itself. If you are up on your code speed, be brave and ask for the 13 wpm test. If you pass this, you can walk out with a General Class license! How come? Because you automatically received credit for element 3.

But wait! Did you do your technical homework thoroughly? Are you feeling pretty confident after two victories? Then ask to take element 4-A, which is required for the Advanced Class license. You skip the code test, as you did in going from Novice to Technician. Pass the written, and you have gone from Novice to Advanced in one visit to the FCC!

You are probably wondering if it is possible to complete the whole route and go from Technician to Amateur Extra at one sitting. It is not only possible but is being done. The applicants for this marathon invariably are former hams who let their original licenses lapse for one reason or another, are bitten by the bug again, do a lot of condensed refresher studying, look up old ham friends, and qualify for the Novice ticket. Armed with this, they are ready for the FCC office.

The code test is given first. If you fail it, out you go, but you can return in a month and try again. If you pass the code but fail the written, all is not lost. When you come back for another try, you do not have to take another code run. The only restriction here is that you must return to the original FCC office.

All licenses are issued by the central FCC records office in Gettysburg, Pennsylvania.

Call Signs

The amateur radio license is a thin piece of paper about the size of a common playing card. One side shows your name and address and the issue and expiration dates. The other side repeats the address and also includes the class of the ticket. Most importantly, it includes your *call sign,* which is your radio identification for as long as you care to keep it. Hams generally refer to this simply as *call.*

The United States and its possessions are divided into ten radio districts, numbered from 1 through 0. The latter is always spoken as zero, never as *oh.* Calls consist of a prefix of one or two letters starting with W, K, N, or A, a district number, and one, two, or three suffix letters. The prefix A is not used singly, but always as the first letter of combinations AA through AZ.

The shortest call is a combination of four characters; the longest, of six. As is the case with automobile tags, short combinations are status symbols. The minimum is one, like W1AW. This is a 1 x 2, the x representing the district number. KH6P is a 2 x 1. NP4ZY is a 2 x 2. So is AL4WJ. K2DUX is a 1 x 3. WB3EXU is a 2 x 3.

Like to have your nickname or initials as part of your ham call? Forget it! Until 1978 only Amateur Extra Class operators could obtain specific combinations, but no more. Extras holding 2 x 3 or 1 x 3 calls still have the privilege of upgrading them to shorter ones, but they have to take what the FCC data bank has available.

From the earliest days of ham licensing, operators who moved from one radio district to another had to apply for new call signs. This was changed in another 1978 ruling. Now your call is yours for life. However, you must inform the FCC of all address changes. If you fail to do so and Gettysburg has one of its letters to you returned marked "Unknown," you're quite likely to have your license revoked. This is happening.

12

Receivers, Transmitters, and Transceivers

For many years, the two basic elements of an amateur station were a receiver, usually with its own built-in AC power unit, and a medium-power CW/voice transmitter of perhaps 100 to 150 watts rating, often with the power pack as a separate unit because of its size, weight, and heat dissipation. This arrangement started to change slowly in the 1950's and 60's with the development of solid-state equipment and the swing from conventional AM modulation to single sideband on the bands from 10 through 160 meters and from AM to FM on the short-range 2-meter band.

Separate receivers and transmitters have been replaced largely by the *transceiver*, which combines their functions in a single cabinet no larger than those of the individual units. Some transceivers are completely solid-state; others use a couple of tubes in the final amplifier stages of the transmit circuits. With few or no filaments to heat, the AC power packs fit right into the chassis, usually with enough space for a small loudspeaker.

A transceiver uses the same frequency-control circuits for both transmitting and receiving. This greatly simplifies normal ham communication. For example: Operator A turns his tuning dial in search of stations calling CQ* on either voice or CW. He hears one, waits for operator B to finish, and then he

---

* See Chapter 16, "How To Be a Good Operator."

12-1. This Heathkit HR-1680 is a representative ham-band receiver for single sideband and CW (code) signals. Takes in the "normal" bands from 10 through 80 meters. Has built-in frequency calibrator.

12-2. Kenwood calls this Model R-300 an "all-band communications receiver" and it is that, with coverage of six frequency ranges from 30 megahertz all the way down to 410 kilohertz. A deluxe set.

immediately presses his mike button or taps his key and calls him. If operator B comes back with a quick acknowledgement, the two are in business. If B answers someone else, A has to look elsewhere on the band, or call CQ himself. If B remains silent after his CQ, A has the option of trying him again. Often this second call is successful.

With a transmitter/receiver combination, operator A would have to tune his transmitter carefully to B's frequency while the latter calls CQ. If B's call happens to be a short one, he might readily be finished before A is ready for him. In the second or two that A needs to touch up his tuning, another operator with a transceiver might catch B's ear first. There's a lot of competition on the ham bands!

Independent units have the one advantage that they permit cross-band operation; that is, transmission and reception on different frequencies. This is necessary when communication is attempted between countries that do not share certain frequencies.

While cross-band capability is not inherent in transceivers, it is an easily added feature. What you need is a small unit called a VFO, for *variable frequency oscillator*. When this is switched on, it takes control of the transmit frequency without affecting the receive. It is such a handy accessory that virtually all manufacturers of transceivers make matching units for their sets.

Receivers

There are two types of shortwave receivers, the *ham band* and the *general coverage*. The application of the first is obvious. The usefulness of the second lies in its very wide frequency range, which includes not only the major ham bands but also CB, domestic and international broadcasting, code practice transmissions, some maritime services, standard time signals, and other surprises. It is a real fun machine for the ham shack. You can put a simple but effective one together yourself in a few evenings from a Heathkit that costs about $100, or spend a lot more for a sophisticated import that has two tuning scales and fifteen knobs on its front panel.

Look in the classified advertising pages of the ham magazines for listings of older tube-type general-coverage receivers bearing such names as Hammarlund, Hallicrafters, or National. These sets have a well-deserved reputation for ruggedness, performance, and longevity, and are good buys if in good operating condition.

Transmitters

Virtually the only ham transmitters on the market come from the few manufacturers who offer ham-band receivers. The units are designed to mate properly for independent or single-frequency operation, break-in keying, automatic switching of the antenna between transmit and receive, etc.

Transceivers

The selection here is very wide. The most popular type covers five or six of the "normal" ham bands from 1.8 megahertz through 28 MHz. Some models do not include the 1.8 MHz segment, as this is close to the upper edge of the regular AM broadcast band and is a possible cause of interference in home receivers of poor design. The 28-MHz band is rather wide, so usually it is represented by three positions of the band change switch: 28 to 28.5 MHz, 28.5 to 29 MHz, and 29 to 29.5 MHz.

A valuable bonus on some sets is a switch position for the single frequency of 15 MHz, for reception of the standard time signals of the world transmitted by stations JJY, in Japan, and WWV, in the United States.

A few sets include AM capability, but this is of little interest because operation on the normal bands is 99 percent single sideband.

A standard feature of transceivers is a "mode" switch. This usually has positions for tune, upper and lower sideband, CW,

12-3. This handsome pair of Kenwood sets consists of the R-599D receiver, on the left, and the matching T-599D transmitter. Provides maximum flexibility on CW, LSB, USB, AM, and FM on receive only. Receiver is all solid-state; transmitter is hybrid.

12-4. The do-it-yourself Heathkit HW-104 is a versatile five-band transceiver with an extra setting for the time-signal station WWV.

and FSK, which means *frequency shift keying*, for radioteletype.

Other circuit features and operating conveniences in modern transceivers are about in proportion to the prices of the equipment. Here are a few to consider.

RIT. This means *receiver incremental tuning*, a means of clearing up signals from misadjusted SSB transceivers. In ham practice the term *round table* describes a net consisting of

12-5 and 12-6. Above: The advantages of digital display of frequency can be added to many transceivers. This example is the Kenwood TS-520S with DG-5 on top of cabinet. Below: Lots of knobs and switches on this Yaesu FT-901D 10–180 meter transceiver enable owner to keep all functions under control. Digital display is self-contained.

three or more stations on the same frequency. If one of them is slightly off tune, his signals in the others' receivers are likely to sound like Donald Duck at his angriest. With RIT, the listening operators can move the receive frequency up and down within a spread of about three kilohertz without affecting the correct transmit frequency, and thus bring the received signals up to normal intelligibility. If only two stations are involved, it sometimes is less trouble for the second station operator to retune his rig to coincide with the setting of the offending station. There's no other choice if neither receiver has RIT.

IF SHIFT. This is a means of moving a signal around in the intermediate-frequency amplifier section of the receive circuitry to reduce interference between wanted and unwanted signals without changing the tuning of the set, which by itself might make matters worse. There is no real way of separating strong signals that are very close in frequency, but in many circumstances IF shift can rescue a signal from a pile of radio-frequency noise.

COMPRESSOR. This is a way of squeezing the operator's voice so as to increase the "talk power" of the transceiver. Sometimes the voice characteristics are changed slightly, but

12-7. A stunning lineup of Yaesu equipment of matching design. From the left: FF-301 loudspeaker, FT-301 transceiver, FV-301 variable frequency oscillator, YO-301 monitor scope. Latter permits analysis of transmitted and received signal wave forms.

12-8. Another Yaesu array, in a different outward design. From the left: loudspeaker, FTd×560 five-band receiver plus WWV, FV400S outboard variable frequency oscillator for cross-band operation.

this would be noticeable only to an operator who is familiar with the other's normal pattern. To a third operator who is a stranger, there would be nothing unusual about the signals.

25 KILOHERTZ CALIBRATOR. This generates accurate frequency markers every 25 kHz up and down all the tuning ranges. It is invaluable, especially for marking the limits of the various ham bands; working out of band is a cardinal sin in ham radio and is very much to be avoided.

MONITOR. Plug in a pair of phones and hear your own voice as it sounds over the air. You might or might not like it! A loudspeaker isn't suitable because acoustic feedback to the microphone usually produces wild howling.

SEPARATE POWER SWITCHES. With hybrid transceivers (those with a few tubes and lots of transistors) it is desirable to turn off the filaments of the tubes during long periods of listening, not only to save AC juice but also to lengthen their lives. This merely requires two switches, one marked POWER and the other HEATERS.

NOISE BLANKER. This helps to reduce noise from sparking devices and machinery such as automobile engines, motors in small home appliances, elevator controls, etc.

CROSS-BAND OPERATION. On a limited but useful scale, this is possible in some of the more advanced transceivers by the addition of accessory crystals in sockets provided for them inside the chassis. There is a choice of fixed receive/variable transmit frequency or variable receive/fixed transmit.

It is also possible to install crystals for normal transceiver operation on single frequencies up to about four, instantly selectable by a switch on the front panel. This feature is great for groups that meet on the air regularly on particular frequencies. There are hundreds of such nets devoted to other hobbies ranging from archaeology through photography and philately to zoology. Listening to them can be amusing and enlightening. If you're lucky, you might be able to talk yourself into one and make many new friends.

COOLING FAN. Transceivers in general and hybrid models in particular are so tightly packed with components that natu-

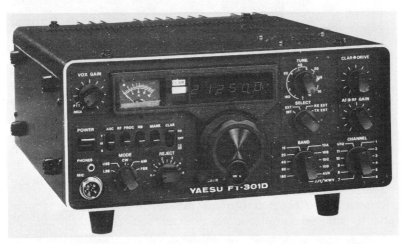

12-9. The Yaesu FT-301D is typical of the new generation of transceivers featuring digital display, wide-band coverage from 10 through 160 meters and JJY/WWV, and SSB, CW, AM, and FSK capabilities.

ral ventilation in the case hardly exists. Thoughtful manufacturers include a noiseless fan about four inches in diameter mounted on the back of the chassis. You don't have to remember to turn it on; it starts when the AC line switch is snapped on.

DIGITAL DISPLAY. No more squinting at a complicated vernier dial and trying to figure out its reading! The big "gimmick" in ham radio is now digital display of the actual tuning frequency in as many as six or seven brightly-lit figures, in a little window over the tuning knob. This feature is original equipment in most top-of-the-line transceivers and is an optional accessory in many other models.

POWER RATINGS. Most transceivers for the normal ham bands are rated as "medium" at 100 or 150 watts. With a good antenna and under favorable circumstances this is all you need to enjoy amateur radio. The FCC maximum is 1,000

12-10. Although the amateur 6-meter band does not seem to be far from either 10 or 2, it behaves differently and usually requires slightly different equipment, like the Kenwood TS-600. This is an all-mode, low-power transceiver just for 6 meters, with the major features of multiband sets except for digital display.

watts (a *full gallon* in ham parlance), a very generous allowance indeed; in most other countries the limit is much lower. A gallon is readily obtained from a large but relatively simple amplifier called a *linear,* which is connected between the transceiver and the antenna.

High power is likely to create some problems, like interference in nearby television, radio, and hi-fi sets, both your own and those of your neighbors. How can you determine if this will happen in your particular location? There's only one way: by a trial run. Unless he is a very good friend, a dealer is not apt to let you take a new linear home on a returnable basis, although he'd certainly feel differently about a used one. A more practical idea is to get in touch with local hams, explain your situation to them, and enlist their aid in borrowing an amplifier from one of them for a couple of days. As the unit is bound to be heavy, at least two men will be needed to carry it. The actual installation won't take more than five minutes, as it consists of nothing more than changing two coaxial cable connections and plugging in the AC cord.

Transceiver Construction Project

A transceiver is a profitable assembly project for a person handy with tools because the end result is a real piece of radio equipment, a whole ham station, that will give him endless hours of enjoyment. Even people with no previous experience whatsoever with a soldering iron have tackled jobs like it successfully.

The particular rig shown here in Figs. 12-10 through 12-19 is the Heathkit HW-101, popular with hams in many parts of the world because it goes together easily. It has many of the features described earlier: coverage of the ham bands from 10 through 80 meters; full SSB operation, with selectable upper and lower bands; CW operation, with a crystal filter for extra sharp reception; provisions for both VOX (pronounced as spelled and meaning *voice operated transmitter* control) and PTT (spelled out and meaning *push-to-talk*); crystal oscillator for frequency calibration of the tuning dial; front panel meter for various tests; etc.

What's interesting about this transceiver from the design standpoint is that it uses 20 vacuum tubes and 19 solid-state devices. This makes it a hybrid, or perhaps a half-breed.

Why so many tubes in this era of the transistor?

The answer is that tube equipment in the form of knock-down kits is much easier to handle, assemble, wire, and repair than their solid-state counterparts. This set is intended for assembly by people with several thumbs working on a kitchen table, not by dainty little Japanese girls glued to binocular microscopes in a surgically-clean factory compound.

The mechanical part of this kit project is strictly nuts-and-bolts. The electrical part is done in two phases. In the first, a lot of components are solder-mounted on nine printed-circuit boards. In the second, after the boards have been fitted to the transceiver's chassis like pieces of a jigsaw puzzle, they are interconnected to each other and to individual parts on the chassis and the front panel by means of an elaborate pre-formed chassis of varicolored wires. This harness saves

12-11. Back view of HW-101 chassis in last stages of assembly. Shielding of high-voltage cage, on the right, will be completed when back plate of cabinet is mounted. Note several sizes of tubes.

12-12. This type of power tube, used in the transceiver, is not generally found in other electronic gear. The cap at the top is the plate terminal. Eight-pin base fits in socket at corner of cage.

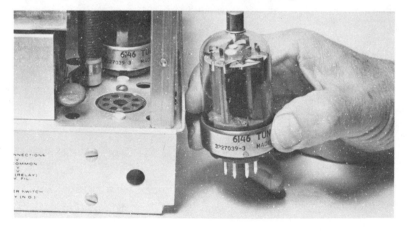

dozens of hours of work over the usual measure-and-cut method of preparing leads.

The power output of the HW-101 is in the medium class:

100 watts on 15 through 80 meters and 80 watts on 10 meters. This is adequate for driving a linear amplifier, if added at any time. The operator can hear his keying through the loudspeaker during CW transmission; this enables him to check his dots and dashes and to correct mistakes as he goes along. For voice transmission any standard dynamic or ceramic microphone can be used.

One of the reasons the chassis is so open and accessible is that no space is taken by an AC power unit. This is a separate unit, which can be placed out of sight behind the cabinet or on the floor under the operating table. It is controlled by a switch on the front panel.

The accompanying illustrations show many of the details of the construction of the set. With the top section of the split cabinet removed, the major components and the small tubes can be reached easily. Two large amplifier tubes and their associated tuning elements are enclosed in a perforated metal cage at the rear right corner of the chassis (as viewed from the rear). Access to them requires the removal of ten screws holding the cover down. This job takes only a few minutes, but it is an effective reminder that the tubes here work on 800 volts DC and that all power to them must be removed *before* a screwdriver touches the screws.

Along the back apron of the chassis are jacks for the connection of an antenna, a CW key, a loudspeaker (an accessory, not included with the kit), a linear amplifier, and a ground wire. There is also a large 11-pin connector plug that takes a mating jack on the end of an 11-wire cable to the outboard AC power unit.

Separate AC Power Pack

This unit is a heavy-duty pack consisting of a line transformer, a batch of solid-state rectifier diodes, four filter capacitors, and a filter choke. On the front of the chassis are a jack to take the other end of the 11-wire cable, a circuit breaker, a line switch, and a bias-adjustment potentiometer.

12-13. The HW-101 is quite deep, as this top view shows. Dark panels are some of the printed-circuit boards; others are underneath. At upper left corner, cage has been left open to show the two power tubes, the triple variable tuning capacitor, and the tuning coils.

The underside of the chassis is a maze of printed-circuit lines, with the wiring harness snaking around them. The small tuning coils for the various bands are covered by a shield plate, with holes in it to permit entry of an alignment tool. This is merely a thin plastic rod with a bit of brass or aluminum stuck in one end to engage the slots of the tuning slugs inside the coils.

The front panel presents a professional appearance. Jacks for earphones and a microphone are in the lower left corner, out of the operator's way. When a headset is plugged in, the external loudspeaker, if used, is cut off. As in most multiband

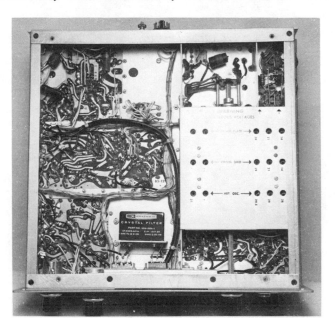

12-14. Underview of the chassis. The light wavy lines in the left center are printed metallic lines imbedded in the circuit boards; they act as connections between small components on the top surface. Band-change switches are under shield plate at right.

transceivers, the 10-meter band is divided into three 500-kilohertz sections between 28 and 30 megahertz for easier tuning. The 15-, 20-, 40-, and 80-meter bands are covered in one revolution of the tuning scale. Band changing is done with an eight-position rotary switch to the left of the big tuning knob in the center of the panel.

A three-position slide switch to the right of the knob gives a choice of VOX, PTT, or calibrate functions. A similar switch further to the right makes the panel meter give readings in three different circuits.

With a proper antenna, tuning the transceiver is a quick and

12-15. Outboard power supply for the transceiver is heavy but compact. Pencil points to important little safety feature: a circuit breaker to protect the equipment and the AC line in case of overloads.

12-16. This is how the transceiver is connected to its AC power unit. Cord from corner of latter goes to the nearest AC outlet.

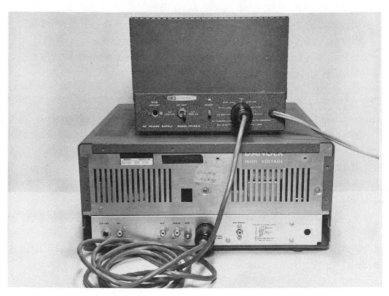

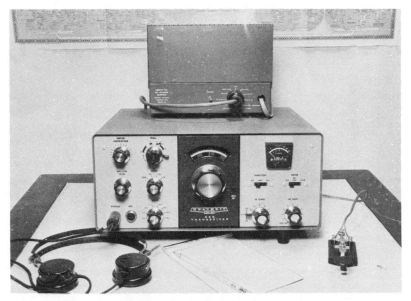

12-17. Above: A whole ham station on a bridge table. Below: A slightly more advanced version, with microphone and matching loudspeaker.

12-18. Surprise! The AC power unit is stowed neatly inside the loudspeaker cabinet, with its controls facing outward for easy adjustment.

simple procedure. In the upper left corner of the panel are the controls that affect the meter movement: mode selector, microphone/CW level, driver/preselector, final amplifier/antenna loading. In the lower right corner are two volume controls and a CW filter switch.

Any small loudspeaker can be used, but a desirable accessory is a Heath speaker in a cabinet of the same design as the transceiver. It is just the right size to also house the AC power unit. Put this speaker on top of the set and you have a complete station that fits on a bridge table with space to spare.

In one of the pictures you will observe that the set sits on the table at an angle, because the front feet are higher than the rear ones. This is done to make it easy for the operator to look at the panel.

Transceiver Accessories

To go on the air with a basic ham rig you need, as a minimum, headphones for reception and a telegraph key for CW (code) transmission. These are never included with the equipment, but are considered accessories. Small loudspeakers are self-contained in some fixed-station transceivers and in all compact transceivers designed for mobile or portable use. Microphones are usually included with the latter two types, in either built-in or hand-held form, but are rarely if ever provided with fixed gear.

All these accessories are available in a profusion of sizes, shapes, and prices.

Headphones   Headphones, earphones, and headsets are interchangeable terms for the head-worn devices that convert audio-voltage signals from a radio receiver into sound. The most widely used type comprises two earphones attached to the ends of an open arc-shaped steel band. Some people prefer a single earpiece, either on a headband or in the form of a small plug of the hearing-aid type that goes into rather than over the ear.

Because early headphones greatly resembled flat tin cans

12-19. With front of cabinet higher than the rear, front panel of transceiver is angled for comfortable viewing by the operator.

and many expensive modern ones still do, hams invariably refer to these sound transducers simply as *a pair of cans*. Not an elegant title, but certainly descriptive. See Figs. 12-20, 12-21, and 12-22.

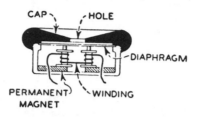

12-20. A cross-section of a bipolar type of headphone.

12-21. Standard type of radio headset, or "cans." Unit on right has been opened to show U-shaped magnet and two windings inside. Diaphragm is thin iron disc, secured by screw-on cap.

12-22. A cross-section of a crystal headphone receiver.

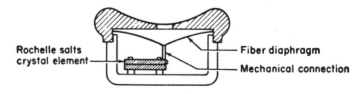

Two distinctly different types of headphones are in use, the *magnetic* and the *crystal*.

The magnetic type consists of a permanent magnet shaped so that it fits into the case. The poles of this magnet are bent up and wound with wire. Phones having only one coil around a central pole are known as *single-pole* phones, as distinguished from the *bipolar* type. The electromagnets are wound with fine insulated wire and the soft-iron diaphragm is held securely in place by means of a screwed-on cap.

When no signal is received, the diaphragm is under a contant pull or attraction exerted by the permanent magnet. When the current increases, the disk moves inward from its neutral or no-signal position. As the current decreases, the pull weakens and the diaphragm moves outward from the neutral position. This back-and-forth vibration sets up sound waves that approximate the voice of the transmitting operator.

The more wire that can be wound on the poles, the stronger the magnetic effect. Space being limited, very thin wire is used. Thin wire has relatively high resistance, so in general the resistance of this type of headphone is an indication of its sensitivity to weak signals. Average figures for two common grades of headphones are 1,000 and 3,000 ohms.

Crystal headphones operate on an entirely different principle. A crystal headphone consists of two piezoelectric crystals (usually Rochelle salt) cemented together to form a single element. This crystal element is mounted in the case of the earpiece with its free end mechanically connected to a fiber diaphragm. Operation depends on the piezoelectric crystals changing their shape when an electric charge is impressed on them. When an alternating voltage is applied to the crystal headphone, the crystal element bends back and forth and causes the diaphragm to vibrate and reproduce the representative sound. Crystal phones are very sensitive over a wide range of audio frequencies and have high impedances.

Telegraph Keys If the army telegrapher who tapped out the news of Custer's Last Stand could be reincarnated as a visitor to a typical ham shack of today, he would instantly recognize something there out of his previous existence: *a telegraph key!* That's how little it has changed in more than a century.

The traditional *straight* key, shown in Fig. 12-25, is nothing more than a spring-loaded switch. Tap on the knob at the front end of the hinged lever and two contacts come together, closing the circuit to which the device is connected. Release your fingers, and the circuit opens. The action is simple, reliable, and utterly foolproof.

There are also two kinds of *semiautomatic* keys, one mechanical and the other electronic. There cannot be a fully automatic one because no machine can think for you.

The mechanical semi uses a vertically pivoted arm, on the right side of which is a small duplicate of the straight key, with the same round knob. This is the dash maker, but instead of tapping down on the knob you tap it sideways from the right. On the left side of the arm is a spring with a contact point on its free end, close to a fixed contact. This is the dot maker. When you press its paddle in slightly from the left, the spring vibrates and starts making dots, up to a maximum of 10 or 12. The number of dots it spews out depends on how long you keep the paddle closed. For the letter E you scarcely touch it; for I and S a light tap is sufficient; for H and 5 another fraction of a second is needed. Be advised in advance; keeping this dotter under control takes practice . . . a great deal of it.

The dot speed is adjustable by means of a sliding weight on the end of the spring assembly. Experienced amateurs can maintain communication up to 35 or 40 words a minute, but they *copy in the head* instead of trying to write that fast, which very few people can do even without the code signals.

A more advanced type of semiautomatic key, more appropriately called a *keyer,* is the electronic. This differs from the vibrator in that it produces both dots and dashes, all perfectly formed and spaced and knob-controllable to speeds in excess of anything any human can copy. It's a great machine, but still no better than the person using it or the one hearing it.

Loudspeakers Loudspeakers follow a more or less standard pattern, as shown in Figs. 12-23 and 12-24. As in an earphone, the central part is a powerful permanent magnet. However, instead of a thin iron disc as in an earphone, the vibrating element that produces sound is a shallow paper cone. The center of the cone is formed into a short cylinder, over which is wound a layer of thin wire; this is called the *voice coil* because the audio signals from the receiver are sent through it. The cylinder fits closely over the pole of the magnet but does not touch it. The outer rim of the cone is cemented to the rim of the stamped sheet-metal frame.

Any current flowing through a coil of wire sets up a magnetic field through and around it. The varying audio signals creates such a varying field in the voice coil, which reacts with the fixed field of the magnet; the cone moves in and out like a piston and sound comes from it.

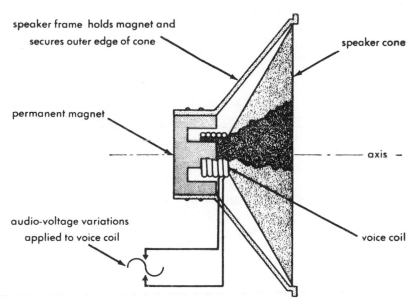

12-23. Basic construction of a permanent-magnet loudspeaker.

12-24. The cone of this speaker has been cut away from its frame to show the position of the voice coil. Connection to the latter is made by two flexible leads sticking out from the paper.

The pitch range of voice is much lower than that of music, so a large loudspeaker of the high fidelity type is neither necessary nor desirable. Most manufacturers offer suitable speakers in cabinets to match the appearance of the sets.

Microphones A microphone converts sound energy into electric energy that varies in frequency and amplitude just as the sound does. The simplest type is the *carbon*, which is used in all commercial telephones. It contains a small cup holding tiny carbon granules. One end of the cup is fixed; the other is attached to a flexible plate or diaphragm. From the central office of the local telephone company, direct current is fed to the mike (common short term) when the handset is picked off its cradle. When the mike is spoken into, the sound waves cause the diaphragm to vibrate, alternately compressing and loosening the carbon granules. This changes their resistance, and the current through them changes accordingly. The steady DC is thus converted into a varying "talking current."

Carbon mikes are cheap and efficient. However, providing them with pure DC from a source other than batteries is a great nuisance, so they are fading from the radio scene. All modern amateur transmitters are designed for self-generating types, as follows.

12-25. "Straight" key favored by most hams for CW transmission. Thumbscrews permit fine adjustment of contact spacing, keying pressure.

The *crystal microphone* works on the principle known as the piezoelectric effect. A varying pressure on a Rochelle salts crystal (or other piezoelectric material) generates a small current that varies with the pressure on the crystal. Crystal microphones are of two types, the *diaphragm* and the *grille*. The first uses a diaphragm mechanically coupled to a crystal element. When sound waves strike the diaphragm, pressure is exerted on the crystals to produce a representative current flow. This type is popular with amateurs for a number of reasons—it is inexpensive, it requires no battery or transformer, and it can be connected directly to an amplifier. The grille type, which has a wider audio-frequency operating range, consists of a group of crystals cemented together in series or series-parallel. Here, sound waves strike the crystals themselves to produce pressure variations.

12-26. Typical dynamic microphone (Electro-Voice Model 664)
found in amateur stations. Heavy
base and rubber feet minimize
creeping on table. "Push-to-talk"
switch on base actuates control
circuits in the transmitter.

12-27. Slender crystal mike can be held in hand or put in table
mount. Center button is on-off control switch for transmitter. In fore-
ground, right, is straight key; in background, vibrator-type semiautomatic key. Operator of this station is prepared for all comers!

Certain ceramic materials exhibit the piezoelectric effect, and are used in mikes in exactly the same manner as crystals. These ceramics are less susceptible to heat and moisture than crystals, and are widely used.

Dynamic microphones are also known as *moving-coil* microphones, because they depend on the movement of a coil in a magnetic field for their operation. A thin coil of aluminum ribbon is attached to a flexible diaphragm to form a unit that moves between the poles of a powerful permanent magnet. Movement of this coil due to sound vibrations causes representative audio voltage to be generated.

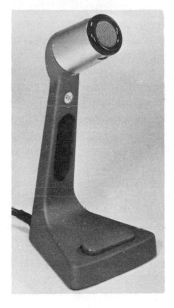

12-28. These are dynamic microphones. Left, table-type, with press-to-talk switch in base. Right, mobile hand-held style, with talk switch at upper left edge. (*Electro-Voice*)

The Cathode-Ray Oscilloscope

The cathode-ray oscilloscope is a versatile and fascinating instrument. It can do such jobs as measuring DC and AC voltages, determining frequency, analyzing complex wave shapes, etc. In a ham shack it is particularly valuable for showing the wave form of transmitted signals. It can also show the quality of received signals. Small scopes especially designed for this application are called *monitors* and are available as accessories from practically all set manufacturers.

Like a vacuum-tube voltmeter, the oscilloscope draws almost no current from the circuit it is measuring. Thus, it can be used directly to measure the output of low-power devices. Its response to rapid AC phenomena is also virtually instantaneous. A VTVM could never follow, for example, the rise and fall of a high-frequency AC voltage. The meter would indicate some mean value because the inertia of the movement is such that it cannot completely follow the high-frequency alternations. The oscilloscope, which employs a weightless electronic beam, can and does follow AC phenomena to 20 MHz and higher.

Because it closely resembles a television set, the oscilloscope is generally regarded a "modern" invention. Curiously, it is one of the earliest of true electronic devices, predating the vacuum tube itself. In almost its present form, it was developed by a German professor, Karl Braun, in about 1897, and even today it is called the Braun tube in some texts. Braun was a pioneer in radio research and made many contributions to the art before he died in Brooklyn in 1918. He never received the popular acclaim that made Marconi's name a household word, but it is significant to note that the 1908 Nobel prize for physics was awarded jointly to him and Marconi, not to the Italian alone.

The heart of the oscilloscope is the cathode-ray tube (CRT), which is similar in design and operation to the television picture tube. As shown in Fig. 12-29, it contains an electron gun, a set of vertical and horizontal deflection plates, and a fluorescent screen. The electron gun consists of an indirectly heated

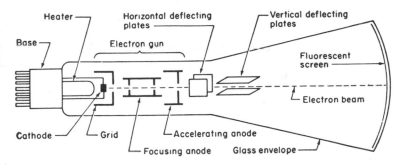

12-29. Basic elements of a cathode-ray tube.

cathode, a control grid, a focusing anode, and a high-voltage or accelerating anode. Just as in a standard vacuum tube, the cathode emits a cloud of electrons. The grid controls the number of electrons that are allowed to escape from the immediate area of the cathode and pass on to the focusing anode attracted by its positive charge.

The grid itself may be biased negatively enough to cut off passage of all electrons to the focusing anode. The focusing anode is equipped with small apertures at each end. Consequently, as the electrons pass through the focusing anode (attracted by the high positive voltage on the accelerating anode), they are made to converge into a stream or beam of electrons traveling toward the screen on the inner face of the tube. The electrostatic fields set up by the focusing and the accelerating anode effectively focus the electrons so that they converge to a point on the screen of the tube (Fig. 12-30A). The screen of a cathode-ray tube is coated with a phosphor compound that emits light (visible on the outer face of the tube) when the electron beam strikes its surface.

Beyond the accelerating anode, the electron beam passes between a pair of horizontal deflection plates and a pair of vertical deflection plates. Since the electron beam is negative, placing a positive potential on any of these deflection plates draws the beam toward the positively charged plate, away from its normal course down the axis of the tube. Placing a negative charge on any of the plates repels the beam from the plate and out of its course down the axis of the tube. Hence

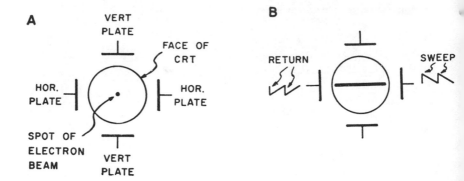

12-30. Displaying an AC sine-wave voltage on the screen of a cathode-ray tube: (A) no deflection voltage, (B) sawtooth voltages applied to the horizontal plates, (C) AC sine-wave voltage applied to the vertical plates, and (D) sawtooth voltage applied to the horizontal plates and AC sine-wave voltage applied to the vertical plates.

the beam can be deflected either horizontally or vertically to move the luminescent spot on the face of the tube either vertically or horizontally from its normal position in the center of the tube. Thus, by applying positive or negative voltages to the horizontal and vertical deflection plates, we can move the spot anywhere on the face of the tube.

If we apply a sawtooth voltage (as shown in Fig. 12-30B) to the horizontal deflection plates, we cause the spot to move very rapidly across the face of the tube from left to right at a linear rate. Notice in Fig. 12-30B that sawtooth voltages of opposite polarity are applied to the two horizontal deflection plates. This is done so that the effect of the voltages applied to the opposing plates will be added. In other words, the voltage on one horizontal plate is pushing while the other is pulling the electron beam. If we were to apply a continuous sequence of sawtooth voltages to the horizontal plates, the spot would move linearly across the screen from left to right, return almost instantaneously, move linearly across the screen, return, and so on, as long as the sawtooth voltages were applied. Thanks to the persistence of the phosphorescence, a horizontal line would appear on the tube's screen. What we have done with the application of the sawtooth voltages to the horizontal deflection plates is to create a linear time base, or sweep. If we now were to apply an AC voltage to the vertical deflection plates (Fig. 12-30C), we have the rise and fall of the

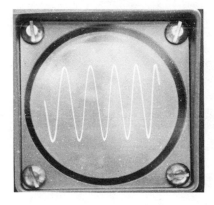
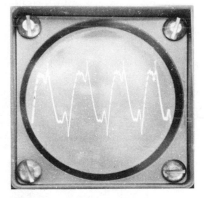

12-31(A). The classical sine wave of alternating current as portrayed neatly on an oscilloscope. Number of displayed cycles depends on frequency of horizontal sweep circuit.

12-31(B). Irregular current through fluorescent lamp is shown here graphically. Input probe was merely touched to metal frame of the fixture. Sharp "spikes" often cause radio/TV interference.

AC voltage plotted against the time required for the spot to travel from left to right on the screen (i.e., the linear time base). The result would be the AC voltage sine wave as shown in Fig. 12-30D. See Fig. 12-31 for photos of actual screen patterns.

The preceding explanation, greatly simplified, shows us how a cathode-ray oscilloscope can be used to give a graphic presentation of varying electrical phenomena. Obviously, this

12-32. This could be any one of dozens of makes of standard 5-inch oscilloscopes. The face of the tube here is covered by a transparent removable screen, which can be graduated to show relative values of imposed signals. The patterns produced by the latter are quite bright and be photographed readily on medium-speed black-and-white film.

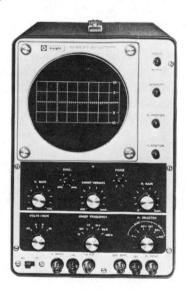

12-33. Inside view of a 3-inch scope made from a kit. Note the neck length of the horizontally mounted picture tube. This section contains the "gun" that shoots electrons at the inner chemically coated surface of the tube.

simple explanation does not tell the whole story. For example, a usable oscilloscope must have circuits for synchronizing the start of the sweep with the beginning of the electrical phenomenon being displayed. Provision must also be made for varying the time it takes for the sweep to travel across the face of the tube if we are to get usable presentations of both low-frequency and high-frequency phenomena.

Oscilloscopes are great fun, and are easy to learn how to use because the slightest change in any adjustment, internally or externally, changes the pattern on the face of the tube. Scopes are available in two general sizes related to the face diameter, 3 and 5 inches, and in both kit and assembled form. They contain no critical elements and are simple construction projects. Three representative scopes are shown in Figs. 12-32, 12-33, and 12-34.

12-34. Small oscilloscope is used to great advantage in this Heathkit Monitor to show actual shapes of sent and received signals.

13

The Antenna Puts Out the Signal

For radio reception over a broad band of frequencies, a wire of almost any length from about 25 to 100 feet, preferably in a straight line, serves quite well. It can be thin or thick, bare or insulated, low or high, preferably the latter. For transmission, however, the length must be related closely to the frequency of the carrier wave. When we convert frequency to wavelength we are able to arrive quickly at some specific figures.

For purposes of calculation, simple formulas are useful:

$$\text{frequency (in hertz)} = \frac{300,000,000}{\text{wavelength (in meters)}}$$

or

$$\text{wavelength (in meters)} = \frac{300,000,000}{\text{frequency (in hertz)}}$$

The figure 300,000,000 represents the speed of radio waves (and also light waves) in meters per second. It can be shortened to 300,000 for frequency values in kilohertz or to 300 for values in megahertz.

A basic resonant antenna for any particular frequency is theoretically half a wavelength long, providing the wire is hanging in free space. However, under practical conditions, the length and the effectiveness of the antenna depend to an unpredictable extent on the proximity of the wire to electrically conductive or reflective surfaces such as tin roofs, attic insulation with aluminum foil vapor barriers, wire lath in walls, etc.; on the height and angle of the wire; and on the presence of supporting wires near the ends of the antenna proper. The element of luck looms large here also. In one location an antenna made strictly in accordance with the formula in Fig. 13-1 will work perfectly the first time the station goes on the air. In another location an identical wire might require shortening or lengthening, or raising or lowering, or a combination of these adjustments, before it puts out proper signals.

Suppose you have qualified for a Novice-Class amateur license, have a transceiver or a transmitter/receiver combination, and are understandably anxious to go on the air with CW to build up operating experience in advance of going for the higher ticket. As a Novice you are restricted to segments of four bands:

28.1 to 28.2 megahertz	("10" meters)
21.1 to 21.2 megahertz	("15" meters)
7.1 to 7.15 megahertz	("40" meters)
3.7 to 3.750 megahertz	("80" meters)

The second band offers sporting opportunities for DX (long-distance) communication, so let's start there. Applying a

$$\text{length of wire (in feet)} = \frac{468}{\text{frequency (in megahertz)}}$$

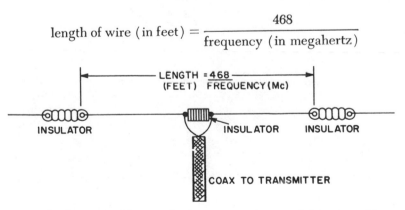

13-1. The basic doublet or dipole antenna is a straight wire with a coaxial-cable transmission line connected in the center.

center frequency of 21.15 to the formula, we obtain the answer of 22 feet plus an inch or so. Allowing some overlap for fastening purposes and experimenting, cut 25 or 27 feet of wire into equal lengths.

Any bare wire will do for the antenna, but your best bet is "7-22," so-called because it consists of seven strands of No. 22 copper wire tightly twisted to form a smooth cable. It is readily available, is easy to handle, is very strong, and is easy to solder.

The Transmission Line

At this point it is important to know that practically all modern ham equipment is designed to connect to the antenna by means of coaxial transmission cable, commonly called *coax* for short. This cable consists of an outer shell of tightly woven

13-2. Installed in a man's den as an experiment, this antenna proved unexpectedly effective for both sending and receiving with a low-power CW transceiver. Center object is connector for the coax feed line.

copper mesh and an inner wire, with the space between them filled with a semihard plastic. The braid is covered with a tough plastic to prevent moisture from seeping in.

By far the most widely used coax is known as RG-8U. It has an outside diameter of $7/16$ inch, and the inner conductor is seven strands of No. 20 wire. This cable can handle the output of the most powerful amateur transmitters. For smaller loads RG-58U coax can be used. This has an outside diameter of $3/16$ inch, with a solid No. 20 in the center.

As protection against lightning, it is important to ground any coax-fed antenna. Just before the point at which the cable enters a house, scrape away a little of the outer covering, connect one end of a heavy wire to the braid, tape the joint well, and connect the other end to a copper grounding rod driven at least three feet into the earth. Nothing, but nothing, can fend off a direct strike, but a properly grounded antenna continually drains off heavy static charges in its vicinity and often does prevent them from building up to the size of a bolt.

Ideally, a transmission line should transfer *all* the RF energy generated in the transmitter to the antenna, without radiating any on its own account. Coax rates very high in this respect because the continuous outer braid acts as an effective shield. There is unavoidably some loss of energy in the resistance of the cable itself and in the insulation between the center conductor and the shield; however, this is relatively slight.

Some hams favor a feeder known as *open-wire*. This consists of two parallel bare wires separated about two inches by thin insulators abut a foot apart. It has very low losses, but it is a notorious snow-catcher, is not as convenient to string as coax, and usually requires an outboard tuner of some kind to match the transmitter properly to the antenna.

Now that you know about coax you can finish the antenna by following Fig. 13-1. At the center insulator connect the inner lead of the coax to one side and the outer shield to the other. This will do for a temporary hookup, but not for a permanent one. There is available a special coax connector which includes the center insulator and a strong clamp for the cable

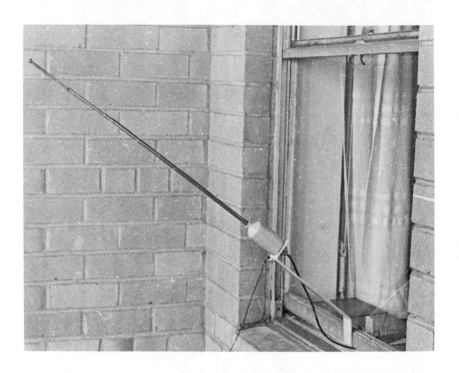

13-3 and 13-4. Intended for portable use, this base-loaded whip is a boon also to apartment dwellers. Above: U-shaped clamp fits neatly under window edge. Right: In the clear-off edge of house eave. Works on seven bands, 2 through 40 meters.

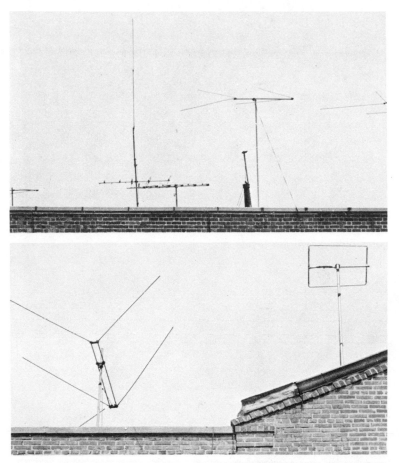

13-5 and 13-6. What to do if you live in an apartment house and the landlord says no to a ham antenna? It's easy to fool him by erecting an inconspicuous one. Top: A simple vertical whip on a TV mast. Above: A CushCraft "Squalo," which is no stranger than the TV antenna on the left. While not as effective as beams, these simple antennas permit much DX operation.

itself. It offers the further advantage that it keeps rain and snow out of the cable.

A straight, simple antenna of this kind is called a *doublet* or a *dipole*. Thousands of hams use nothing more complicated,

13-7. Above and below left: No climbing! Popular fold-over tower is easily raised or lowered with hand winches. All work on antenna made safely at ground level.

13-8. Right: Graceful antenna up 40 feet.

and enjoy highly successful DX. Its major shortcoming is that it is limited to one band. However, there is nothing to prevent you from erecting similar wires for other bands. The required materials are very cheap and it takes only a minute to swap connectors at the transmitter. For another small investment you can add a coaxial switch, which eliminates even this small bother.

You may want to try 40 and 80 meters. If you pick 7.12 MHz as a good spot, the antenna comes to 65 feet plus about eight inches. This is still an easy length to accommodate in most backyards or even on the roofs of apartment houses. For 3.72 MHz, the length figures to 125 feet 10 inches. A few minor bends in this stretch, to make the wire fit convenient trees, chimneys, or other supports, don't seem to do any harm.

In any case, the length of the coaxial *feeder* or *transmission line* is not critical. In private homes it is not likely to exceed 50 or 60 feet, and in most apartments probably 100 feet. Long runs of as much as 200 feet in tall apartment houses are not uncommon.

Beam Antennas

Horizontal antennas tend to radiate energy more strongly in the directions at right angles to their length than off their ends. The effect is often distorted by reflections of the signal from trees, metal-foil insulation in the walls or ceiling of a house, nearby power lines, etc. However, it does exist, and it is made useful in *beam* antennas.

13-9. Reflector and director elements, added to a dipole antenna, increase the radiation in the direction shown.

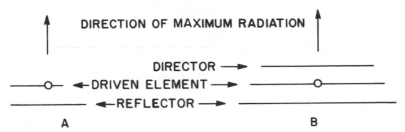

If a second dipole slightly longer than the basic radiating antenna is placed parallel to and a critical fraction of a wavelength behind the latter, it acts as a reflector and makes the radiator, or *driven element*, push more of its energy toward the front than toward the back. See Fig. 13-9A. The reflector doesn't have to be connected to this element; it acts almost in an optical manner. This is not surprising, since radio waves, especially at high frequencies, have many of the characteristics of light waves.

Except for a special type of beam called the *cubical quad*, beams do not use wire elements. Instead, they are made of aluminum tubing from about 1 to 1½ inches in diameter.

A two-element beam is appreciably better than a bare dipole, is light and therefore easy to erect, is less obtrusive than most television antennas, and is popular with hams who live in houses with limited roof area.

The signals going out forward from the driven element can be enhanced further by the addition of a *director* element in front of it, as in Fig. 13-9B. The action of this element might be compared to that of a concentrating lens in front of a light.

As more reflectors and directors are added, the beam effect of the antenna increases, and more and more of the energy goes forward at the expense of reduced radiation to the rear. However, both the length and the spacing of the extra elements become very fussy at the same time. The back radiation is never really eliminated in practical antennas used on the popular frequencies, but the *difference* between the front and back signals is very noticeable. Sometimes you can tell from an operator's remarks that his beam is facing 180 degrees away from your location, yet his signals are loud and clear; this is one of the vagaries of shortwave transmission. However, if he should turn the antenna to face you, the loudspeaker is likely to chatter in protest at the increased volume.

Because beam antennas must be mounted on some sort of swivel or rotator, for aiming purposes, they tend to be rather complicated physical structures. It is impractical to use full doublets on 40 and 80 meters, because these would require large-diameter tubing 65 and 125 feet long. For these bands,

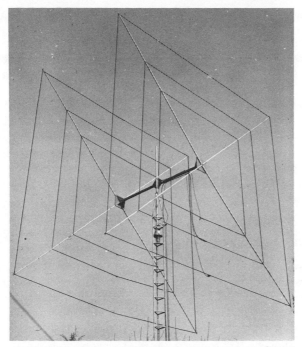

13-10. Bird catcher? No, signal getter and sender. A two-element "quad" made of wires on X-shaped frames. Works on 10, 15, and 20 meters and is highly directional. As many as six sections can be mounted on a long horizontal boom.

when beams are used at all, the practice is to use much shorter elements and to lengthen them artificially by the insertion of simple loading coils of heavy wire. For 20 meters and less, the problem resolves itself, because the elements are naturally shorter.

Maximum performance on any one band is obtained from an antenna dimensioned for that band alone. However, at the cost of a little loss of overall efficiency it is possible to make both plain dipoles and multielement beams work on several bands with one feed line from the transmitter. The trick is to break up the elements with *traps*, so-called because one of their functions is to confine particular frequencies to portions of the elements. Their other purpose is to act as loading coils, to make the elements think that they are longer electrically than they are physically. The positioning of these traps is very critical.

By far the most widely used and successful antenna of this type is a three-element beam for 10, 15, and 20 meters, with one driven element, one reflector, and one director, each with a tuned trap in each half of its length. There are endless variations of this construction, and an inquisitive amateur can spend years experimenting with them.

Raising the Antenna

Stringing a wire dipole is a fairly simple job. Raising a beam takes much more planning and work. The ideal arrangement

13-11. Modest 15-foot tower with 10-foot mast supports fine assortment of antennas. At top is 2-meter vertical; center, 6-element Yagi (named for Japanese designer) for 2 meters; bottom, three-element trapped beam for 10, 15, and 20 meters. Flat roof made installation job safe and easy. Base of tower rests on wood platform that distributes the load over the roof surface.

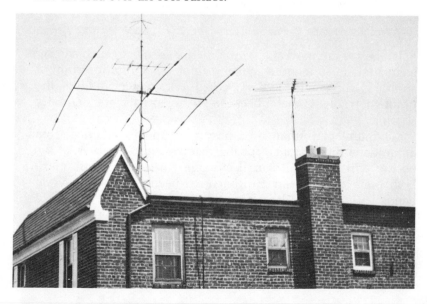

is a three-leg, open-lattice iron or aluminum tower, with one or two telescoping sections that provide a maximum height above ground of 40 to 60 feet. The tower must be of the fold-over type, so that the retracted structure can be lowered to a horizontal position to make the antenna itself and its rotator motor accessible for inspection, repair, experimentation, etc. A poured concrete base two or three feet square and about five or six feet into the ground is needed to give the tower a reliable footing.

In most installations of this type the actual support for the tower is not the base but a very heavy iron pipe from 3 to 6 inches in diameter and from 13 to 18 feet long, stuck in the concrete to about half its length. At the top is a large hinge, welded in place. This is joined to a plate on one side of the tower and secured with husky ½-inch bolts.

Raising the tower from the horizontal position or lowering it, and winding the telescoping sections up or down, are done with surprising ease by means of geared-down hand winches. One is attached to the post and by means of strong steel cable

13-12. Connected between a transmitter and almost any length of antenna wire, this little matching coupler usually works wonders in coaxing signals out of the combination. Unit measures only about 5 × 3 × 6 inches.

controls the angle of the tower. A second is attached to one side of the latter and by means of a second cable does the second job. When the tower is vertical it is secured at its bottom by a rod through a U-shaped bracket welded to the bottom of the post.

For hams who have no muscle but plenty of money there are motor-operated winches that do all the work at the touch of a switch. You can always tell when a ham in the neighborhood has installed the lazy man's antenna. Whenever he has visitors he shows them how the thing goes up and down!

A somewhat simpler, cheaper, but shorter fold-over tower is made of aluminum, is hinged at its bottom to a flat ground plate or a shallow concrete block, and is intended for mounting alongside a house. At about the first-floor level it is secured to a hook or eye bolt in the wall by a piece of wire.

13-13. An ideal antenna for a small house and lot. The tower rises to 43 feet, telescopes to 20, and folds over. Antenna itself is a three-element, three-band beam for 10, 15, and 20 meters, with a short vertical rod at the top for 2 meters. Cables on side of house connect to beam and its rotator (dark area under antenna boom).

13-14. How *not* to site a ham tower. Note the power-line pole at the left, with cross-arms carrying high-voltage lines in two directions. If the tower ever comes down in a storm there will be fireworks!

When it must be lowered, the latter is replaced by a length of rope looped through the hook (better yet, through an ordinary clothes-line pulley). While one man nudges the tower away from the wall, another keeps a firm hold on the rope and lets the structure down slowly. To raise it, the procedure is merely reversed.

For people who don't have the space for towers there is a cheap and simple alternative for putting an antenna in the air: common TV masts. These are available in straight 5- and 10-foot sections that join end-to-end and in telescoping models that go as high as 36 feet. Installation is made easy by a wide variety of mounting hardware. Beyond about 15 feet the assembly must be guyed. The obvious disadvantage here is that the guys must be loosened when the antenna is to be lowered. This isn't as onerous as it sounds, because the mast is relatively light and the guy wires or ropes themselves can be used to ease the structure down and haul it back up.

A word of incidental advice. Before considering a ham tower, check with the local housing authorities and make sure there are no zoning restrictions against them.

The Unfriendly Landlord

What can you do if you live in a rented apartment and the landlord refuses—even for money—to allow a ham antenna of any kind on the roof of the building?

Take heart. Hams have lived with this problem since the days of crystal detectors and high-button shoes, and they lick it by using what they like to call an *invisible* antenna. Well, almost invisible. More accurately, it is a random-length wire tacked or taped through several rooms, usually along the baseboards for concealment. It should avoid direct contact with all water, gas, and steam pipes and air ducts, as these are invariably well grounded.

Any thin wire, insulated or bare, is satisfactory. You might be able to salvage some from the electromagnets of a defunct

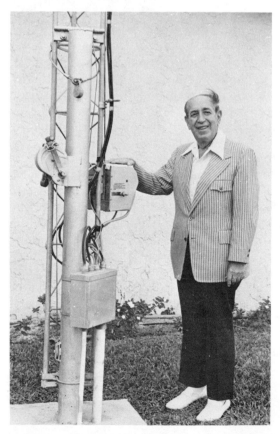

13-15. Monroe Freedman, W4MGL, does not have to crank his 60-foot tower up and down by hand. He has a motorized winch (in box) that does the job in minutes. Cables to house are in buried pipes. Concrete foundation of tower is six feet deep.

door bell or from a transformer in an old radio or TV set. Or it might be easier to buy a 100-foot roll of No. 22 insulated wire from a radio supply store for less than two dollars. Starting from the table that holds your ham gear, unspool the wire as far as you can in one direction through the apartment. Don't cut it; you almost certainly will try another configuration that calls for a longer run.

Remove about an inch of insulation from the set end of the wire and push it into the antenna connector on the back. Turn on the rig for receive only. *Do not attempt to transmit at this time.* Unless your apartment is completely encased in a shield of metal lath, you will surely hear stations immediately; in fact, lots of stations, because radio waves have an astonishing ability to pass through ordinary building materials.

From the critical transmitting viewpoint the operation of a

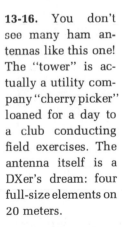

13-16. You don't see many ham antennas like this one! The "tower" is actually a utility company "cherry picker" loaned for a day to a club conducting field exercises. The antenna itself is a DXer's dream: four full-size elements on 20 meters.

random-length antenna is quite different from that of the idealized beams described earlier in this chapter. A transmission feeder such as coax is not used; the entire length of the wire is active antenna. Actually, this antenna is a "Marconi," with the grounded set acting as the ground element. (See Chapter 1, describing Marconi's experiments.) Any attempt to use it for transmission is almost sure to cause serious damage in the transmitter.

Fortunately, this trouble can be avoided by the use of a rather simple device called a *transmatch* between the transmitter and the antenna. See Fig. 13-11. As its name implies, it converts the unknown characteristics of the random wire to the established characteristics of the transmitter. With a transmatch, you will know after about two minutes of twirling its dials whether your invisible aerial is putting out signals. In fact, you'll probably work some stations right away!

Continue to experiment. Try working on several bands. Most transmatches cover the wide range of frequencies from 1.8 to 30 megahertz. Of course, you have to stay within the operating limitations of your license.

You may find it both amusing and instructive to try the transmatch with a variety of metallic objects such as the following: unaltered TV and FM aerials; car antennas; clothes lines with a thin center wire; metal rain gutters on a house; or almost anything made of aluminum, such as porch furniture, window and door frames, screens, and flag poles. Even an opened umbrella.

Believe it or not, all of these have been used successfully for both reception and transmission!

A transmatch is virtually a necessity when a temporary station is set up in a tent or other shelter during field day exercises, which are popular outdoor activities for ham clubs throughout the country. Beams and their supports are hardly transportable, so the usual field antenna is a piece of wire thrown up a tree. This requires careful matching with equipment designed primarily to work with coax.

Since almost any high wire works better than any low one, the organizers of field days usually canvass the club mem-

13-17. "Out of this world!" This is what visitors usually say about this ham antenna. Literally, they are right. It is a 27-foot parabolic dish built by Vic Michael, W3SDZ, and used by him for bouncing signals off the moon. Vic owns broadcast station WMLP, so he leads a full radio life. He makes most of his own amateur equipment.

bership and try to find a ham who is also an archery enthusiast. With a hunting bow, he can easily loft a string-carrying arrow to a height of 50 feet; the string then is used to pull up the antenna wire itself. Fishing line generally serves as the string.

Vertical Antennas

Where space in a horizontal direction is limited or a roof is not readily accessible, amateurs resort to vertical antennas. They lack the directional characteristic and the effect of in-

13-18. Unlike the towers that support beam antennas, this one is actually part of the antenna itself. The topmost section is a slender length of aluminum tubing, adjustable in length for tuning purposes. Because the structure offers very little wind resistance it does not require guying in most locations. Vertical antennas of this type are popular because they are relatively cheap, are light in weight, and easy to erect.

creased power offered by beams, but if adjusted properly they punch out good signals. In its simplest form the vertical is generally a piece of aluminum tubing an inch or more in diameter, mounted on but insulated from a stiff pipe driven into the ground; or it might be mounted against the side of a house with U-shaped clamps. It is not usually guyed, because it has relatively little wind resistance and can withstand considerable swaying without snapping.

A vertical is essentially a quarter-wave antenna, the other quarter section being provided by the ground itself. Its length depends on the frequencies to be covered. An 18- or 20-foot piece, for example, with a small loading coil connected to its base, radiates well on 40 and 80 meters, depending on how much of the coil is used. With the judicious use of traps at various points, some verticals work on three, four, or five bands.

The Dummy Antenna

Preliminary tuning of a transmitter should be done with a *dummy* antenna rather than with the regular outdoor one, to minimize interference with stations already on the air. A dummy is merely a large noninductive resistor of about 50 ohms, which is about the equivalent value of a properly-sized antenna. Some hams try to use ordinary 50-watt lamps for this purpose, but they are not satisfactory because their resistance changes with the temperature of their filaments. See Figs. 13-18 and 13-19.

13-19 and 13-20. Above: Typical hookup of a dummy antenna (black can) with a transceiver and a coax switch (next to can). This back view shows routing of cables. Left: Heart of dummy is resistor inside protective tube, attached to lid of gallon can filled with oil.

14

144 Megahertz and Up

In the 1970's there was a spontaneous move from AM (amplitude modulation) to FM (frequency modulation) on the amateur 2-meter band; that is, 144 to 148 megahertz. This occurred because large quantities of high-grade commercial FM tranceivers had become surplus at bargain prices as a result of changes in FCC regulations. The equipment worked slightly off the scale from 2 meters but could be revamped easily for ham purposes, so it was gobbled up quickly.

At first, hams used these sets in conventional manner: *simplex*, between home stations, between the latter and mobile sets in cars, and car-to-car. Simplex means that two stations in contact use the same frequency.

Amateurs knew from previous experience with 2-meter AM that this band is good primarily for line-of-sight communication, so they weren't surprised to learn that the mere change in the method of modulation didn't change the situation. Mobile-to-mobile operation particularly was spotty, as intervening structures often blocked the signals completely. FM ham radio on 144 megahertz would have died aborning if not for the rebirth during the very same period of an unrelated technology: the *repeater*, first used on the ham bands as far back as the early 1930's.

A repeater is a combination of a receiver and a transmitter with an omnidirectional antenna atop a high building. The prefix *omni* means that the antenna functions uniformly in all directions. Because in effect it can look down on and can be seen by small antennas on the street below, it greatly extends the reliable talking range of mobile transceivers and hand-carried flea-power "walkie talkies." With this antenna advantage, repeater transmitters don't need to be very powerful. Ratings run between about 35 and 100 watts.

The secret of successful repeater operation is its use of dif-

14-1. Below: Mounted on an inverted U-shaped dash bracket, with microphone quickly available in a spring clip, this Heathkit 2-meter transceiver works directly off the car's battery.

14-2. Bottom: When not needed for the car, the same unit makes a fine little home station, working off the AC power supply at the left. Sound issues from loudspeaker in bottom of transceiver cabinet.

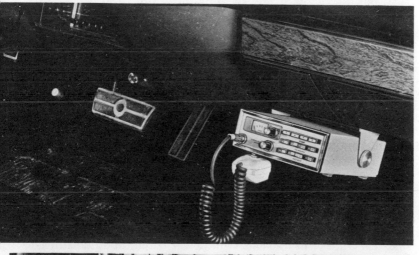

ferent frequencies for reception and transmission (this is called *duplex*), and of highly sophisticated control devices spun off from the computer field. In its simplest form the routine is as follows:

Operator A on one frequency presses his microphone button once but remains silent. The short carrier wave from his transmitter is picked up by the repeater receiver and in turn is fed into the station's control circuit, which automatically turns on the repeater transmitter. Almost instantly, Operator A hears a rushing sound in his receiver, which means that the repeater carrier is on the air. Quickly, he squeezes the mike button, holds its down, and proceeds to call Operator B. If the latter is listening on the repeater transmit frequency, he waits for A to stop talking, waits a few seconds more until the carrier rush disappears, and then takes control of the carrier himself by squeezing his mike button.

Why doesn't B answer A immediately after A says "over"? He could. However, practically all repeaters have automatic timers that limit each one-way transmission to a few minutes, so if he talked too much he might readily find himself talking to himself.

"Crazy, man, crazy!" is the comment made by most hams when they hear these goings-on for the first time. They are also impressed by the dead silence of FM in the absence of ac-

14-3. This compact Yaesu "Memorizer" offers digital readout of 800 FM channels between 144 and 148 MHz. Takes only 2.5 amps at 13 volts (car battery) in transmit, with power output of 10 watts. Note front foot. This raises front edge of set, and gives better sound projection from the bottom-mounted loudspeaker.

14-4. Above: See that little box with the buttons, on seat next to Monroe Freedman, W4MGL? It's a "Touch Tone" encoder, by means of which he can make land-line phone calls through some 2-meter repeaters. Great, isn't it? **14-5.** Right: Radio is family affair here. Wife Rosalyn is W4MOU. Note call-letter license plates.

tual signals, the result of effective *squelch* circuitry. There is none of the static, hash, heterodyne whistles, etc., that often bedevil communication on AM and SSB.

Repeater operation is quite different from conventional ham practice in another important respect: it is *channelized*, according to gentlemen's agreements worked out by repeater

These advanced transceivers use plug-in modules.

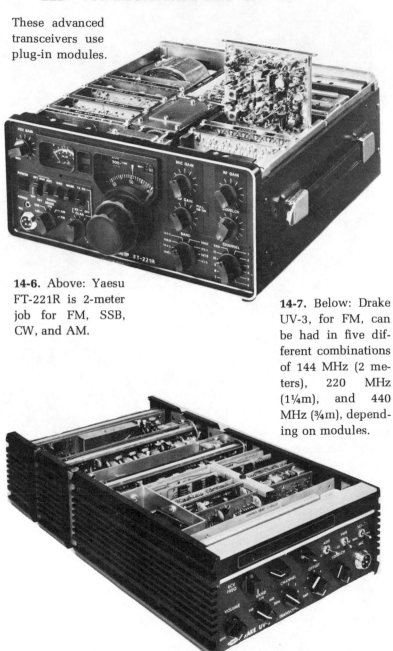

14-6. Above: Yaesu FT-221R is 2-meter job for FM, SSB, CW, and AM.

14-7. Below: Drake UV-3, for FM, can be had in five different combinations of 144 MHz (2 meters), 220 MHz (1¼m), and 440 MHz (¾m), depending on modules.

14-8. Slung from the shoulder like a camera, the Icom IC-502 is a 2-meter single sideband portable with full 3 watts of power. Works on internal "C" cells or car battery; good for field exercises and the like.

groups all over the country. You don't just twirl the tuning dial of your rig and listen for an open niche; you must know the receive/send frequencies assigned to particular repeaters in your area. These are called *pairs* and are usually identified only by the two digits following 146 or 147; the first is the receiver *input* (your signal), the second the transmitter *output* (the signal you hear). Between 146.010 and 146.970 megahertz the input signal is 600 kilohertz lower than the output. Between 147.000 and 147.990 the input is 600 kilohertz higher than the output. Thus, 34/94 indicates input of 146.340 megahertz, output of 146.940 megahertz; 93/33 means 147.930 in, 147.330 out. The assignments in the two halves of the 146-148 part of the 2-meter band are staggered to minimize interchannel interference and to avoid duplication of numbers. (In the megahertz figures, the final 0 is often left off, as a matter of convenience).

The frequency of 146.52 is reserved nationally for straightforward simplex operation, which is devoted mainly to DX and experimental work. Amateurs with radioteletype facilities chatter together on 10/70 duplex.

Most repeaters are operated by ham clubs, but there are also many run by individual amateurs. Most are open to anyone in the vicinity, but some can be *keyed* only by club members. So many have sprung into existence that you can drive from coast to coast and be within talking distance of one about 90 percent of the time. In fact, so many are active on 2 meters that in some areas newcomers are going to 220 and 420 megahertz instead.

14-9. Left: Similar in appearance to the Model IC-502, this Icom IC-215 is for FM. It offers a total of 15 channels. **14-10.** Right: The Icom IC-22S is of the popular cigar-box shape, suitable for both mobile and fixed station use. Offers 22 channels of FM. Nominal power output 10 watts, with a low setting of one watt.

14-11. The Kenwood TS-700S is an all-mode transceiver that puts you on 2-meter SSB, FM, CW, or AM at the flick of a switch. Box at right is an outboard variable frequency oscillator, useful for experimental purposes.

14-12. Top: Work SSB and CW on 144–145 MHz and FM on 146–147 with this Icom IC-211 fixed-station transceiver. Continuous digital readout of frequencies. **14-13.** Right: A really handy idea. Heathkit microphone combined with telephone encoder. **14-14.** Below: Like the IC-22S, this Icom IC-245 is a 10-watt FM transceiver, with the added feature of digital frequency readout. An optional accessory is an SSB/CW adapter.

Equipment Roundup

Equipment for all three bands is plentiful, attractive, versatile, compact, but not especially cheap. Practically all of it is of the transceiver type. Some models include CW, AM, and SSB in addition to basic FM. With rigs like these you can be busy on the air most of the day and night! The pictures herewith give some idea of what the market offers.

14-15. The Tempo VHF/One Plus, shown here with an SSB adapter, is a relatively high-power FM transceiver of 25 watts. Full digital read-out, 800 selectable frequencies controlled by buttons on the microphone.

14-16. Kenwood TR-7400A is an unusually flexible FM 25-watt transceiver. All functions are represented by well-marked controls. Any of 800 channels set up in seconds and appear in read-out window. Bottom unit is AC power pack.

14-17. The section of pipe just above the window shade is the simplest form of vertical antenna for 2 meters. With a 10-watt transmitter it readily "accesses" repeaters 20 miles distant.

14-18. Hams in more remote locations might not find a simple whip antenna enough for good communication on 2-meter FM. They can do much better with a multielement beam like this Cush-Craft "Yagi." The active driven element is the heavy one, near the vertical pole. The element between it and the pole is a reflector; the others to the right are directors.

15

Setting Up the Ham Shack

As you listen on the amateur bands you will always hear hams refer to their operating quarters as "shacks." By dictionary definition a shack is a small house or cabin that is crudely built and furnished. Obviously this description does not apply to any of the stations pictured throughout this book. Many of them are elegantly panelled rooms, and the furnishings, meaning the ham gear, can cost several thousand dollars. So why the appellation *shack*?

The term dates back to the earliest days of wireless, when its first use was in ship-to-shore communication. Existing vessels had no suitable cabin space for the equipment and the operator, so as these became available a little house was built for them on an upper deck as close as possible to the bridge. House is not a nautical word but cabin is, and from cabin eventually evolved shack.

Station Layout

In Chapter 12 there is a picture of a complete amateur station laid out comfortably on an ordinary bridge table. This shows what can be done when floor space in a house or apartment is limited. The bridge table will serve the purpose quite nicely until the owner decides, as he most surely will, to acquire additional equipment such as an amplifier for the transceiver, a monitor scope, or an antenna analyzer. Dozens of useful and

15-1. This spacious operating table in the basement shack of Paul Hertzberg, K2DUX, is a common door, sanded smooth and varnished. The support is made of 2 × 3″ lumber, with a shelf to hold magazines, call books, etc. Note the wall plastered with acknowledgement cards from stations all over the world that Paul has worked with Heath SB-110, at left, and Collins "S" Line, two units at right.

interesting ham accessories are available, and it is indeed a strong-minded amateur who can resist the temptation to buy them as his finances permit.

The accompanying pictures show a number of typical stations ranging from the very simple to the very elaborate. A newcomer can study them with profit, for they show arrangements that he might be able to adapt to his own circumstances.

In an Apartment

The first place that suggests itself as a ham shack in an apartment is a bedroom that is either unfurnished or is used as a spare room for visitors. The latter won't mind a table in the corner loaded with interesting gadgets, and they might even want to play with them themselves.

The best operating table to hold the gear is a desk between four and six feet long, with comfortable knee space and either one or two stacks of drawers. Desks for office use generally are made of steel; for home use, of wood in more decorative designs. Look in the yellow pages of the telephone directory for dealers in used or reconditioned furniture. Their prices are usually low and you can bargain with them, especially if you can take the piece home yourself.

Also consider the possibility of assembled or partially assembled but unfinished desks and matching chairs. They are

15-2. Then and now! Photos below and top of facing page illustrate the remarkable technical development of ham radio over the years. Below: The 1934 cellar shack of W2DJJ, with all home-brewed equipment. Receivers at left; no loudspeaker, but note large "cans" on center set. Open breadboard transmitter is at right, with most parts salvaged from broadcast receivers of the period.

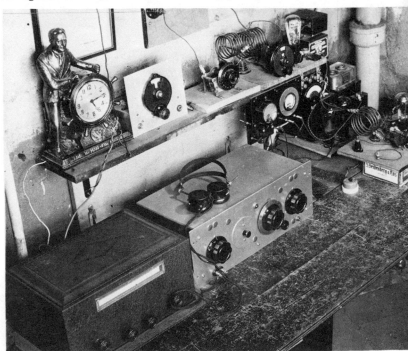

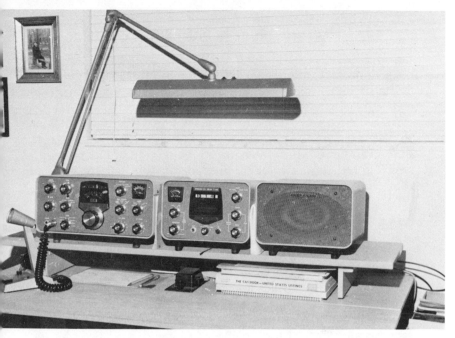

15-3. The 1978 shack of the former W2DJJ, now K4JBI. Nothing hay-wire here! The only wire in sight is the coiled cord of the micro-phone, at the left. All equipment is also home built, but from Heath kits: transceiver at left; combined clock, phone patch, antenna meter in center; and loudspeaker with AC power supply, right. Short shelf creates convenient storage area below.

sold in sanded-smooth condition and take liquid or spray paints and enamels very well.

The last place you might think of as suitable for a ham shack is the kitchen, but why not if a corner of a breakfast nook can be spared? A 2-meter transceiver with its AC power pack needs no more shelf space than a double toaster, and many normal-band rigs are smaller than a breadbox.

Faced with a space problem in a small apartment, some ham families tackle it by treating ham gear as decoration rather than as technical equipment. With its graceful cabinet, its array of knobs and dials, and its blinking lights, a modern transceiver has definite eye appeal and is invariably a conversation piece—very literally a conversation piece!—for first-time visitors. Its effect is enhanced by an illuminated globe on the table or perhaps a large framed map of the world behind it.

As a last resort, think of a closet whose contents can be stowed elsewhere, perhaps in shallow boxes under a bed. These can be found in the housewares section of many department stores. The author of this book once helped a friend shoe-horn a transceiver, a linear amplifier, a loudspeaker, a

15-4. Two five-foot desks in the spare bedroom of their Florida home hold the dual stations of Monroe and Rosalyn Freedman, who are W4MGL and W4MOU, respectively. Below: Monroe's main rig is the Collins "S" Line, with a linear amplifier on the floor behind his chair. Roz's personal transceiver (facing page) is an Atlas Model 210. The boards that hold a large assortment of auxiliary equipment are ready-cut and finished book shelves, free-standing on two sizes of screw-on table legs. This ingenious arrangement permits quick and easy shifting of the equipment. One of the certificates on the wall of this den is the Golden Anniversary Award of the Quarter Century Wireless Association to Monte for his fifty years of amateur service.

power supply, a monitor scope, and a beam direction-control box into a clothes closet with a floor area measuring only 26 by 53 inches! There was even room overhead for books, magazines, and assorted supplies on one shelf.

Since the door of the closet interfered with the lone chair for the operator, it was removed and replaced by an attractive roll-up screen. This shack too became a conversation piece, with the limitation that only one person at a time could examine it.

In a House

The advantages of a house are plain: it usually has more rooms than an apartment and certainly more room. That is, besides the normal kitchen, dining room, living room, and bedrooms, it is also likely to have a sun parlor or patio, a basement, an attic, and a garage.

15-5. Some hams enjoy working at the controls of several layers of equipment. Not Sol Weingast, WA2WXT, of Atlantic Beach, L.I. He prefers this tidy arrangement of his Collins receiver/transmitter combination, with the end units angled slightly inward for easy viewing.

15-6. Two tables in L formation provide a comfortable operating position for Nathaniel Pfeffer, W2AIM, New York, N.Y. All equipment in this shack is in open view. Room was originally a den.

15-7. With the exception of a VTVM, an oscilloscope, and a teletype-writer sender, all the equipment in this interesting shack was hand-made by the owner. He is DL6AW, Horst Rohmann, Braunschweig, Germany. A technician employed by the well-known Rolleiflex camera works, he is accustomed to precision construction.

15-8. The attractive living room of his home in Bamberg, West Germany, is also the radio shack of Hannes Bauer, DL1DX. With a call sign like that he should and does work DX with his Drake gear.

What these five shacks have in common is their Drake equipment, which is American-made. However, the layouts differ greatly and reflect the habits and operating practices of their owners. **15-9.** Top of this page: Ed Cobb, K4YFR, Norfolk, Va., has the sets in an open U arrangement, with all controls within arm's length. **15-10.** Bottom: Tom Softley, W3ZSX, Laurel, Md., has mike on flexible mount, at mouth level. **15-11.** Top, next page: Earl Smith, K2YLM, Eatontown, N.J., evidently likes a busy two-level table. **15-12.** Center: Ray Grenier, K9KHW, Milwaukee, Wis., prefers a neat table, and lines up his extra equipment on two shelves above it. **15-13.** Bottom: Mark Franklin, WBØANT, Minneapolis, Minn., favors the straight-line design. Note large 24-hour clock in upper left corner of picture.

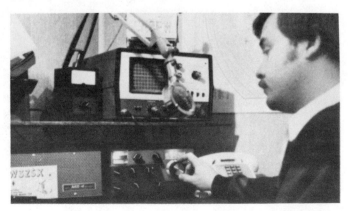

15-14. What does a top executive of the world's largest retailer of consumer electronic equipment pursue as a hobby? Golf? Poker? Fishing? If you figure that he gets enough of electronics during the day, the last leisure activity you'd be inclined to mention is ham radio. However, that's precisely the long-time preference of Lawrence Blostein, K5BJ, national advertising and sales promotion manager of the Radio Shack chain, shown above in his own shack!

As a high-school student in Chicago just before World War II, Larry was W9BUD, with a one-tube CW transmitter and a two-tube regenerative receiver—a far cry from his present Yaesu FT-101B transceiver and matching VFO, Realistic DX-160 for general shortwave listening, and incidental accessories. Glad to grab an experienced brasspounder, the Air Corps put him to work as a radio operator at bases in the United States and overseas. Out of uniform, Larry was graduated from the University of Illinois with a degree in journalism and a major in advertising, landed a job with Allied Radio, the predecessor of Radio Shack, and continues to combine business with pleasure in Fort Worth, Texas, his present QTH.

Of these latter facilities, the basement is surely the best bet for a comfortable ham shack. If part or all of it is already finished off as a den or playroom you can move in with a desk or table and all your equipment and be ready to connect the antenna fifteen minutes later. If it is unfinished you can easily partition off part of it, preferably near a window so that you experiment conveniently with antenna feed lines.

Take into account the need for heat in the winter and possible air-conditioning in the summer. Depending of course on your latitude, you may be able to get by during cool weather with a portable radiant electric heater and during warm weather with an exhaust fan.

If the cellar beams are exposed, cover them with squares of ceiling tile, which is not tile at all but is made of light acoustic material. These will help to confine noise from your loudspeaker; for this, the rest of the family upstairs will be grateful.

A sun parlor, with two or three walls made mostly of windows, is a cheerful location for a shack, providing it is well insulated and has the same heating and cooling facilities as the rest of the house. In many of the southern states an open patio, screened or not, is habitable for three or four months of the winter but generally is impossible during the summer, when the sun temperature can rise beyond 100 degrees.

A corner of a clean garage is also good, especially if it is convenient to a workbench and its tools and if its temperature can be controlled.

North or south, the attic of the average single-family house can be made livable only by the addition of costly insulation in its roof and the extension of electric lines and heating pipes or ducts from rooms below; in the summer, a large air-conditioner is the only machine that will maintain life there.

16 ————————————

How to Be a
Good Operator

Once you begin operating a station, remember that you are just one of many amateurs using the ham frequencies. You will find the bands quite crowded. Respect the rights of your fellow amateurs. *Before going on the air, check with your receiver to see if anyone else is using the frequency on which you wish to transmit. If someone is on it, wait until he is finished or change to another frequency.*

CW (Code) Transmission

In communicating by means of code, it is important to follow certain procedures. Before you actually begin transmission, it is a good idea to listen to other amateurs until you are familiar with the established practices.

1. *Calling a station:* When calling a specific station, send its call letters four or five times, then the letters DE (the French word for *from*) followed by your own call letters sent four or five times, and end with $\overline{AR}$ (end of message). Thus, a call might be as follows: W4XYZ W4XYZ W4XYZ W4XYZ DE W4ABC W4ABC W4ABC W4ABC $\overline{AR}$. Signals with an overline are sent as one letter. After making your call, wait an appreciable interval before calling a second time.

2. *General inquiry call:* When you wish to have any amateur who hears your signal answer your call, send out a general inquiry call. This is done by sending the letters CQ four

or five times, then the letters DE followed by your call letters and the letter K (go ahead). Thus, a general inquiry call might be as follows: CQ CQ CQ CQ DE W4ABC W4ABC W4ABC W4ABC K. After making a general inquiry call, tune your receiver very slowly two or three times over the frequency band on which you are operating to pick up any possible answers. To avoid an excessive number of answers, make your CQ specific. For example, if you wish to call anybody in Connecticut, you should send as follows: CQ CONN CQ CONN CQ CONN CQ CONN DE W4ABC W4ABC W4ABC W4ABC K.

16-1. A neat, uncluttered layout of equipment promotes good operating habits. Note comfortable position of operator's arm as he uses semiautomatic key. QSL cards from all over the world testify to his skill.

You will probably find that your results will be better when you specify a particular locality in your CQ calls.

3. *Answering a call:* When answering either a direct or a general inquiry call, send the call letters of the calling station two or three times, DE, and your own call letters followed by K. For example, an answer might be as follows: W4XYZ W4XYZ W4XYZ W4XYZ DE W4ABC W4ABC K.

4. *DX calls:* The phrase DX is used when calling foreign countries. For example, a general inquiry call for foreign countries would be as follows: CQ DX CQ DX CQ DX CQ DX DE W4ABC W4ABC W4ABC W4ABC K.

5. *Discouraging break-ins.* After a call to a particular station, or at the end of each transmission during a contact already in progress, use the ending signal $\overline{KN}$. This means that you do not want anyone else to break in.

6. *End of contact.* When you're finished saying everything you wanted to say to the other chap, but intend to stay on the air for contacts with other stations, sign off with the signal $\overline{SK}$. For example: $\overline{SK}$ W4XYZ DE W4ABC. This is an invitation for others to try.

7. *Closing down.* When you're finished with a station and intend to close down immediately, add CL after your call sign: $\overline{SK}$ W4XYZ DE W4ABC CL, and turn the switches on your equipment to *off*.

The Q Code and Other Abbreviations

To speed up code transmission many amateurs make use of the Q code and the commonly used abbreviations that follow. Whether or not you use these short cuts, it is a good idea to become familiar with them so you will not be at a loss when you find amateurs who do.

Q Signals These are three-letter groups, originally formulated during Marconi's time to expedite the exchange of vital nautical information to and from ships at sea. When a Q signal is followed by a question mark (.. __ ..), a question is being asked; when it stands alone, it is translated as affirmation or reply.

The complete list consists of 65 signals, of which about three dozen have applications in ham CW operating practice. For voice operation some of them have assumed the meaning of nouns and verbs and are fixed parts of ham vocabulary. For example:

QRM. *Interference.* "I'm being QRM-ed by WF6XPH."

QRN. *Static.* "Can't read you through the QRN."

QSL. *Acknowledgement.* "Would appreciate a QSL card from you."

QSO. *Communication.* "Just had a long QSO with OA4J in Lima."

QTH. *Location.* "The QTH here is Houston, Texas."

Signal	As a Question	As a Reply
QRB	How far distant are you?	My distance is . . .
QRG	What is my frequency?	Your frequency is . . .
QRH	Is my frequency steady?	Your frequency is steady.
QRI	How is my tone?	Your tone changes.
QRJ	Are my signals weak?	Your signal is weak.
QRK	Are my signals legible?	Legibility is (1 to 5).
QRL	Are you free to handle traffic?	I am busy now.
QRM	Are you meeting interference?	I am being interfered with.
QRN	Are atmospherics bothering you?	Atmospherics are bothering me.
QRO	Shall I increase power?	Increase your power.
QRP	Shall I use less power?	Decrease your power.
QRQ	Shall I send faster?	Send faster.
QRS	Shall I send slower?	Send slower.
QRT	Shall I stop sending?	Stop sending.
QRU	Have you any messages for me?	No traffic for you.
QRV	Are you ready?	I am ready.
QRW	Shall I notify . . . that you are calling him?	Notify . . . I am calling him.
QRX	Shall I stand by?	Stand by until I call you.
QRZ	Who is calling me?	You are being called by . . .
QSA	What is my signal strength (1 to 5)?	Your signal strength is (1 to 5).
QSB	Does my signal strength vary?	Your signal strength varies.

QSD	Is my keying correct? Are my signals distinct?	Your keying is incorrect; your signals are bad.
QSG	Shall I send . . . telegrams at a time?	Send . . . telegrams at a time.
QSK	Shall I continue?	Continue with traffic.
QSL	Can you give me acknowledgment of receipt?	I give you acknowledgment of receipt.
QSM	Shall I repeat last message?	Repeat last message.
QSO	Can you communicate with . . . direct?	I can communicate with . . . direct.
QSP	Will you relay to . . . ?	I will relay to . . .
QSV	Shall I send v's?	Send a series of v's.
QSW	Will you send on . . . kHz with . . . type transmission?	I will send on . . . kHz with . . . type emission.
QSX	Will you listen for . . . on . . . kHz?	I will listen for . . . on . . . kHz.
QSY	Shall I change to . . . kHz?	Change to . . . kHz.
QSZ	Shall I duplicate each word?	Duplicate each word.
QTC	How many messages have you?	I have . . . messages.
QTH	What is your position in longitude and latitude?	My position is . . . longitude and . . . latitude.
QTR	What time is it?	Exact time is . . .

Abbreviations for CW Work Abbreviations help to cut down unnecessary transmission. However, make it a rule not to abbreviate unnecessarily when working an operator of unknown experience.

AA	All after	OM	Old man
AB	All before	OP-OPR	Operator
ABT	About	OSC	Oscillator
ADR	Address	OT	Old timer
AGN	Again	PSE	Please
ANT	Antenna	PWR	Power
BCI	Broadcast interference	PX	Press
		R	Received

BCL	Broadcast listener	REF	Refer to; referring to; reference
BK	Break		
BN	All between; been	RPT	Repeat; I repeat
B4	Before	SED	Said
C	Yes	SEZ	Says
CL	I am closing my station	SIG	Signature; signal
		SINE	Operator's personal initials or nickname
CLD-			
CLG	Called; calling		
CUD	Could	SKED	Schedule
CUL	See you later	TFC	Traffic
CW	Continuous wave	TMW	Tomorrow
DX	Distance	TNX	Thanks
ECO	Electron-coupled oscillator	TT	That
		TU	Thank you
FB	Fine business; excellent	TVI	Television inter-ference
GB	Good-bye	TXT	Text
GE	Good evening	UR-URS	Your; yours
GG	Going	VFO	Variable-frequency oscillator
GM	Good morning		
GN	Good night	VY	Very
GND	Ground	WA	Word after
GUD	Good	WB	Word before
HI	The telegraphic laugh	WD-WDS	Word; words
		WKD-	
HR	Here	WKG	Worked; working
HV	Have	WL	Well; will
HW	How	WUD	Would
LID	A poor operator	WX	Weather
MSG	Message; prefix to radiogram	XMTR	Transmitter
		XTAL	Crystal
N	No	XYL	Wife
NIL	Nothing; I have nothing for you	YL	Young lady
		73	Best regards
NR	Number		

Signal Classification

Upon establishing contact either by voice or code, the first thing that most amateurs do is exchange information on the quality of their signals. To facilitate this exchange, the RST system shown below is commonly used.

READABILITY (R):	1—unreadable
	2—just readable
	3—readable with difficulty
	4—readable
	5—exceptionally readable
SIGNAL STRENGTH (S):	1—very faint
	2—very weak
	3—weak
	4—fair
	5—fairly good
	6—good
	7—quite strong
	8—strong
	9—exceptionally strong
TONE (T): (applies only to CW)	1—hissing note
	2—rough AC note, no tone
	3—rough AC note, slight tone
	4—AC note, fair tone
	5—varying tone
	6—varying tone with some whistle
	7—slightly varying tone
	8—very slight varying tone
	9—constant tone

With voice transmission, there is obviously the opportunity for a more or less detailed explanation of respective signal quality. With code, however, the RST system provides a quick and fairly exact means of telling one another the quality of each other's signal. When a contact "pounds out" that your transmission is R4 S5 T7, you have a good idea of how he is receiving you.

Station Identification

When you get into an interesting QSO it is sometimes easy to forget that you are in effect talking on a huge radio party-line, not on a private circuit, and that you have to let other listeners know just who you are in order to avoid confusion. The FCC

16-2. When this picture was made of Wayne Tope, WB4TUP, of Riverview, Florida, in 1975, he was twenty-two years old, had been a licensed ham for four years, and was the recipient of four national awards for public service. He was the first person outside of Nicaragua to know of a disastrous earthquake that hit the country a few years ago, and he handled more than a hundred important messages dealing with the relief efforts. By the time this new edition is published Wayne will have been graduated from a theological seminary, with a degree in psychology and will continue his public service as a full-fledged minister and active ham. Wayne is definitely what you'd call a good operator!

regulations in this respect are very clear. The following is a pertinent extract from paragraph 97.87 of the rule book:

> An amateur station shall be identified by the transmission of its call sign at the beginning and end of each single transmission or exchange of transmissions and at intervals not to exceed 10 minutes during any single transmission or exchange of transmissions of more than 10 minutes duration. Additionally, at the end of an exchange of telegraphy (other than teleprinter) or telephony transmissions between amateur stations, the call sign (or the generally accepted network identifier) shall be given for the station, or for at least one group of the stations, with which communication was established.

To make it painless to obey this rule, manufacturers of ham equipment offer a convenient accessory: a timer that turns on a bright red light or a loud buzzer for a few seconds at 10-minute intervals. It can also be found as part of station control "consoles" that also include a 24-hour digital clock, a phone patch, an antenna meter, and other goodies.

Keeping a Station Log

The FCC regulations state that every amateur must maintain a station log containing the following information: (1) date and time of each and every transmission, (2) station calling and station called, (3) frequency of operation, (4) power input to the transmitter final stage, and (5) closing time of contact and the name of the operator. Any suitable ruled book can be used as a log. The FCC has no objections if the log contains remarks on signal quality, personal comments, notes, etc., in addition to the required information. A number of excellent commercially printed amateur log books are available.

The FCC requires that your station's log be retained for a

period of at least one year following the date of the last entry, and be made available to the FCC upon request.

Voice Procedure

In voice operation, the same sequence of call letters is used. The intermediate signal is now "this is" or "from," usually the former, and the ending signal is "over" or "go ahead," again usually the former. The expression "come in please" is an invention of movie-script writers, and is never used by any self-respecting operator.

It is contrary to FCC regulations to use the reverse form: "This is WA2XX calling WB3ZZ."

The rules state further: "The Commission encourages the use of a nationally or internationally recognized standard phonetic alphabet as an aid for correct telephone identification." Unfortunately, it doesn't specify any one alphabet, and leaves the choice up to the individual operator. This has led to a proliferation of wild combinations of words and letters, some amusing and many not at all helpful. For years the headquarters staff of the American Radio Relay League (ARRL), the national ham organization, recommended a simple alphabet consisting mostly of unmistakable first names, and this was quite successful. However, in the 1970's it switched to another list adopted by the International Civil Aeronautics Organization (ICAO). This offers no advantage over the previous one, as you can tell by scanning the two.

PHONETIC ALPHABETS

ARRL	ICAO	ARRL	ICAO
A—Adam	Alfa	F—Frank	Foxtrot
B—Baker	Bravo	G—George	Golf
C—Charlie	Charlie	H—Henry	Hotel
D—David	Delta	I—Ida	India
E—Edward	Echo	J—John	Juliett

K—King	Kilo	S—Susan	Sierra
L—Lewis	Lima	T—Thomas	Tango
M—Mary	Mike	U—Union	Uniform
N—Nancy	November	V—Victor	Victor
O—Otto	Oscar	W—William	Whiskey
P—Peter	Papa	X—X-Ray	X-Ray
Q—Queen	Quebec	Y—Young	Yankee
R—Robert	Romeo	Z—Zebra	Zulu

The letter Q in particular seems to have been a mistake. It is supposed to be pronounced as "kay-beck" in the French fashion, but this means absolutely nothing to English-speaking operators, most of whom stick to the more recognizable "Queen." Also, six of the new ICAO words have three syllables replacing perfectly good words with only two, hardly an improvement.

One word that isn't on either list but is heard more

16-3. Enormous advantage of single sideband equipment is its virtual freedom from TVI (television interference). With TV receiver directly alongside ham equipment and TV and ham antennas on same mast, the owner of this station can operate any time without bothering the screen.

frequently than any is "Roger," which represented the letter R in the military systems of the World War II period. Since R in International Code is the signal of receipt meaning, "Okay, I heard you," the word Roger naturally was used for the same purpose in voice communication. Although R became Robert in the ARRL list and then Romeo in the ICAO, neither ever replaced Roger in its original meaning. Use Romeo instead of Roger the first time you're on the air with A3 and the reply is likely to be, "Oh, yeah? And I'm Juliett."

Occasionally you might hear an operator use "Kay" at the end of a transmission. This is an obscure word referring to one of the legendary knights of the Round Table, according to Webster's New World Dictionary. After you hear it several times, you will realize that it is a phony phonetic for the letter K, which is International Code for "go ahead."

First Time on A3

You've been monitoring the ham bands, you've obtained your General Class license, and you're ready to go on the air with voice for the first time. There's no reason to be nervous, because A3 is pretty much like talking on the telephone. Here's a little scenario that might help you overcome what little "mike fright" you might suffer for a few minutes. We'll use a couple of imaginary call signs.

"Hello CQ, CQ, CQ, CQ, CQ, any 20-meter phone, CQ, CQ, CQ, this is WJ6PXH, Whiskey Juliett number 6 Papa X-Ray Hotel. CQ, CQ, calling any 20-meter station, this is WJ6PXH, WJ6PXH, Whiskey Juliett number 6 Papa X-Ray Hotel. What say someone please? Over."

You release the mike button, and luck is with you.

"WJ6PXH, WJ6PXH, this is WF4LK, Whiskey Foxtrot number 4 Lima Kilo, WF4LK. Do you read me, old man? Over."

"Yes, yes. WF4LK from WJ6PXH. Thanks for the shout. The handle here is Bill, the QTH Azusa, California. Alpha Zulu

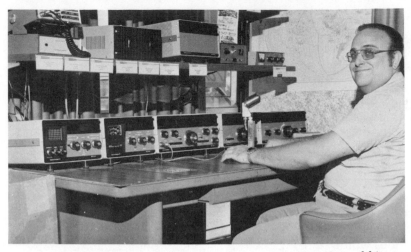

16-4. Who wouldn't be smiling and happy with this all-Heathkit station built around the SB-104 transceiver? Owner is Sherman Leifer, who lives in Florida and whose call sign is W4FLA. Previously, he lived in the eighth radio district and his call was, fittingly, K8HAM.

Uniform Sierra Alpha. You're putting in a very good signal. Would appreciate a report on mine, as this is my first QSO on A3. WF4LK from WJ6PXH. Over."

(Note that the reply from WF4LK is much shorter than your CQ, because he knows that anyone calling a simple CQ will surely listen on his own frequency and will quickly pick up his own call when he hears it. Your own comeback is also brief because you know that WF4LK has tuned you in properly.)

"WJ6PXH from WF4LK. Roger, Roger. Handle is Henry and I'm in Arlington, Virginia, just outside of Washington. Well, you sound just great. What are you using there?"

(Note here that Henry hasn't used an ending signal. Communication is good and the interrogatory tone of the last sentence obviously calls for an answer, so Henry merely releases his mike switch, his carrier goes off the air, and you can answer him.)

"I'm running an Atlas 250 barefoot. Hold on a minute, Henry, my phone's ringing."

"Roger."

(Note here that the formalities are dropped in these quick exchanges).

"WF4LK, this is WJ5PXH. Henry, I have to pick up my wife at the supermarket. Thanks again for the report. Hope to see you again. 73. WF4LK from WJ6PXH, pulling switches."

"So long Bill. WJ6PXH, this is WF4LK, clear."

In ham lingo, *handle* means *name, running barefoot* means you are using a medium-power transmitter or transceiver, without a linear amplifier to boost its strength, and 73 means *best regards.*

Be careful not to fall into a habit common to new operators: the repetition of everything the other chap said in his last transmission.

"Roger, Roger, old boy, on your Yaesu 101 and linear and your Mosely three-bander on that 60-foot mast."

He knows what he has and he just described it to you, so why this senseless feedback?

Using a Phone Patch

You will often hear calls like this:

"CQ, CQ, CQ, Miami Beach for a phone patch, CQ, CQ, Miami Beach for a phone patch. This is WL2DXU, Whiskey Lima number two Delta X-Ray Uniform in New York. Over."

This means that WL2DXU wants a ham in Miami Beach to buzz a local number and to put the answering party through his transmitter and in turn into WL2's receiver, so that they can carry on direct communication. A *phone patch* at the Miami Beach station makes this possible. It is a rather simple accessory and is often part of a station operating console, as mentioned in the section of this chapter entitled "Station Identification."

Be very careful in using a phone patch. Before you offer to

provide it, get clear answers to some questions from the calling ham:

"Is this a local call? If it's out of my area, can I make it collect? Does the other party know you, and by what name? Has he or she ever been on a patch before? Are you calling for yourself or for another person in your shack? Are you aware that the call must only be personal and that you must not mention business matters of any kind?"

The fourth question is particularly important, because if the answer is no you might have a very sticky time trying to explain what it's all about.

In spite of all this, patching can be a rewarding service if it is performed for service men and women overseas and for civilians in foreign countries, all of whom want to talk to relatives back home. But not all countries have "third-party" agreements with the United States, so be very wary about handling patches in either direction. For current data on the situation, check the news items in the monthly ham publications.

Interference

Every amateur must make sure that his transmitter is free of the defects that result in interference with other radio transmission. If your transmitter is operating on the assigned amateur frequencies and is properly tuned, the only interference you are likely to cause is with commercial radio and television broadcasts. In regard to this the FCC Regulations, Part 97 (Amateur Radio Service), state the following:

> 97.131 *Restricted operation.* (a) If the operation of an amateur station causes general interference to the reception of transmissions from stations operating in the domestic broadcast service when receivers of good engineering design including adequate selectivity characteristics are used to receive such transmissions and this fact is made known to the amateur station licensee, the amateur station shall not be operated during the hours from 8 P.M. to 10:30 P.M.,

local time, and on Sunday for the additional period from 10:30 A.M. until 1 P.M., local time, upon the frequency or frequencies used when the interference is created.

Thus, it is clear that when your transmitter interferes with your neighbor's radio or television reception, you should change to a frequency that does not cause interference or go off the air during the specified "silent periods." In consideration of your neighbor's listening or viewing pleasures, you will probably stay off the air for additional periods. Notice that the FCC regulation states that the necessity for self-imposed silence is dependent upon the affected radio and television receivers being of "good engineering design" and possessing "adequate selectivity characteristics." In many cases this is a legal way for you to avoid the specified silent periods, since the fault often lies with the affected receiver and not your transmitter. However, even if you can prove to your neighbor that his radio or television set, not your transmitter, is the real culprit, you are sure to bring about a strained relationship when you persist in interfering with his radio and television reception. When you learn that your transmitter is causing interference, the wisest course is to shut down when your neighbors are using their sets, until you can correct the cause of the interference or determine those amateur frequencies that do not cause interference.

The following discussion on broadcast and television interference will help you understand how much interference is caused and how it often can be eliminated. However, if the suggested remedies do not correct the situation, you should, if possible, enlist the aid of a more experienced amateur in solving your particular problem. The job is fascinating and you will increase your knowledge of electronics.

Broadcast Interference (BCI)

Interference with the reception of AM broadcasts does not often occur since the superheterodyne receiver with a loop antenna, generally used today, is fairly selective and capable of rejecting all but the strongest random noise signals outside its operating range of 550 to 1,650 kHz. On the other hand, if a broadcast receiver is located close to your transmitter or uses an outside antenna that is erected near your transmitting antenna, it can be affected by your transmitter. Here are some of the common types of BCI, with suggestions for their elimination.

Key Clicks Key clicks, resulting from a keying envelope that is too square, are short pulses of radio-wave energy throughout the radio spectrum. Fortunately, the energy contained in the frequencies above and below the transmitter's frequency is small. However, key clicks will be picked up by receivers in the immediate vicinity of the transmitter. Using a filter to round out the keying pattern will eliminate this disturbance.

Sparking at the key contacts creates a disturbance similar to that created by electric-light switches and motors. This type of keying disturbance can be distinguished from that actually radiated from the transmitter by disconnecting the antenna and using a dummy load. Key-contact sparking can be reduced by a filter.

Overmodulation More than 100 percent modulation with phone transmission generates disturbances similar to key clicks. Overmodulation, of course, should be avoided for other than interference reasons.

Saturation Saturation or blocking of a nearby broadcast receiver might occur when the receiver uses an outside antenna near your transmitting antenna. With the owner's permission, the receiver's antenna might be shortened. Also the transmitter's and the receiver's antennas should be placed at right angles to one another to minimize coupling.

Receiver Faults Often the receiver itself is at fault. For example, the transmitter frequency might be such that when mixed with a harmonic of the receiver's local oscillator, the IF frequency results. In this case the broadcast receiver near your transmitter would receive your transmission loud and clear. If your signal does not interfere with a broadcast station, your neighbor should not object. However, if your signal conflicts with his listening, change your frequency to one that does not disturb him.

Occasionally, your voice transmissions will be heard over the entire band of a nearby broadcast receiver. This usually happens because of a rectifying action in one of the receiver's early stages. If your transmitter is close, the signal is naturally strong. This can sometimes be corrected by installing a wave trap (parallel tuned circuit in series with the receiver antenna). When the wave trap is tuned to your transmitting frequency, the high impedance it offers effectively keeps the signal from the faulty receiver's input.

Overall Interference   If necessary, the effects of all forms of broadcast interference can be reduced by screening the transmitter and installing power-line and antenna filters as described for TVI in the following paragraphs.

Television Interference (TVI)

Interference with television presents a much more difficult problem. The major cause of practically all TVI can be traced to the harmonic frequencies that are radiated by both the amateur transmitter and its antenna. Since television Channels 2 through 6 operate from 54 to 88 MHz and Channels 7 through 13 from 174 to 216 MHz even- and odd-order harmonics from all amateur bands below 30 MHz (28, 21, 14, 7, 3.5, and 1.8 MHz) fall within the television frequencies. Harmonics generated on the 50 MHz amateur band also fall in the upper range of television frequencies from 175 to 216 MHz. Harmonics generated in the 7-, 3.5-, and 1.8-MHz amateur bands present little problem, since only harmonics of the eighth order and above are present in the range of television frequencies. (The eighth-order harmonic of 7 MHz is 56 MHz, which is just inside the lower range of television frequencies.) Harmonic frequencies of such a high order from a transmitter are usually so weak that they only affect a television receiver in the immediate vicinity of the transmitter.

However, with harmonics generated in the 28-, 21-, and 14-MHz amateur bands, the situation is different. Here, for example, the second harmonic of 28 MHz is 56 MHz, which falls in the frequency band used by Channel 2.* Since harmonics generated by a fairly powerful transmitter up to the sixth order are capable of interfering with surrounding television reception, the TVI problem that can arise when transmitting on the 28-, 21-, and 14-MHz amateur bands is readily apparent. (The problem also arises on the 50-MHz band, where the fourth harmonic falls into the upper television band.)

You may well ask, then, how TVI is to be avoided when transmitting on these critical bands. The easiest solution, of course, is to stay off these critical bands during viewing hours if surrounding television receivers are affected. However, this curtailment of operating time can be avoided if you can reduce the amount of harmonics radiated from your transmitter and antenna to an ineffective level. Notice that the word used is "radiated" and not "generated." By their nature, oscillators generate harmonics of the carrier frequency. These harmonics have only a fraction of the energy contained in the carrier frequency and are quickly attenuated as they radiate away from the transmitter. Nevertheless, they are still strong enough to affect a sensitive television receiver, especially in a fringe area of television reception where the television signal is itself very weak. Another important point about TVI is the wattage rating of the amateur transmitter. Obviously, a 1,000-watt transmitter will cause more TVI than a 50-watt transmitter, since its harmonic output is correspondingly stronger.

To what extent can you QRP (that is, reduce transmitter power) without finding yourself off the air altogether? To an

---

* See Appendix C for the frequencies of the various television channels.

incredible extent, meaning that you can go from 1,000 watts all the way down to one watt and still work a lot of stations! The only catch is that you need a good antenna, but that's a requirement for high-power work too.

As an example of what hams are accomplishing on flea-power, consider the record rolled up in less than two years by Ron Moorefield, W8ILC, of Dayton, Ohio. Using a Ten-Tec Argonaut transceiver and a four-element monoband beam on 20 meters, he has worked other amateurs in 180 countries—all with one watt—and he has QSL cards to prove it. When the editor of a ham magazine expressed some polite skepticism about his claims, Ron promptly sent him photostatic copies of 100 of the cards.

In modern transmitters direct radiation of energy is mini-mized by the use of low-impedance capacitors across meter and jack leads, the power-transformer primary, and numerous other circuit elements, and by thorough shielding of the chassis itself. Radio-frequency coils are often enclosed in indi-vidual shield cans, and larger assemblies are placed inside cages made of "hardware cloth" (perforated sheet metal) or of wire mesh. Finally, the entire chassis is mounted in a sturdy metal cabinet with a minimum of openings through which RF signals might escape.

It is easy to determine whether TVI is due to radiation from the transmitter or to the major radiation from the antenna. Dis-connect the latter and hook the transmitter instead to a dummy antenna; operate the rig on various frequencies and look for interference on your own television receiver.

It is practically impossible to eliminate *all* harmonic radia-tion, because it is the natural result of oscillator action. In many cases, a simple cure for the TVI is a tuned filter con-nected between the aerial lead and the aerial binding posts of the affected receiver. This usually traps out the weak remain-ing harmonic signals and restores peace in the household. Such filters are cheap, take only a minute to install, and do not affect normal television reception.

It is interesting to note that transmitters of the SSB type are very easy to "de-bug" of TVI, for the simple reason that the carrier and one sideband are suppressed and only one sideband gets to the antenna. With approximately two-thirds of the possible sources of harmonics eliminated, the remaining third is readily cleaned up. A low-powered SSB rig, properly tuned and coupled to a proper antenna by a nonradiating coaxial transmission line, can usually be operated directly alongside a television receiver without making any noticeable impression on the latter. This is one of the reasons why hams have switched over from conventional double-sideband AM to single-sideband AM.

17

Frequency, Time, and Distance

Electric power companies, telephone companies, radio and television stations, navigators of ships and planes, and also the amateur radio fraternity, all depend heavily on precise frequency and time information. They must have a constantly available source—a reliable, nationally and internationally recognized *standard*—with which to compare and regulate their own timing equipment. For over 50 years, the National Bureau of Standards (NBS) has been providing this standard for most users in the United States.

Since the inception of the broadcast services from radio station WWV in 1923, NBS has continually improved and expanded its time and frequency dissemination services to meet the ever growing needs of an ever widening community of users. Today, still striving for better ways to serve its public, NBS is making major contributions to the nation's space and defense programs, to worldwide transportation and communications, and to a multitude of industrial operations, as well as providing convenient, highly accurate time service to many thousands of individual users around the world. Services are presently available from stations WWV and WWVB in Fort Collins, Colorado, and from WWVH in Kauai, Hawaii. In addition, new calibration services using network television are also available. See Fig. 17-1.

NBS broadcasts continuous signals from its high-frequency radio stations WWV and WWVH. The radio frequencies used are 2.5, 5, 10, and 15 MHz. All frequencies carry the same program, but because of changes in ionospheric conditions, which sometimes adversely affect the signal transmissions, most receivers are not able to pick up the signal on all frequencies at all times in all locations. Except during times of severe magnetic disturbances, however—which make all radio transmissions almost impossible—listeners should be able to receive the signal on at least one of the broadcast frequencies. As a general rule, frequencies above 10 MHz provide the best daytime reception while the lower frequencies are best for nighttime reception.

Services provided by these stations include: time announcements, standard time intervals, standard frequencies, propagation forecasts, geophysical alerts, marine storm warnings, UT1 time corrections, BCD time code.

Fig. 17-2 gives the hourly broadcast schedules of these services along with station location, radiated power, and details of the modulation.

The time and frequency broadcasts are controlled by the primary NBS Frequency Standard in Boulder, Colorado. The frequencies as transmitted are accurate to within one part in 100 billion at all times. Deviations are normally less than one part in 1,000 billion from day to day. However, changes in the propagation medium (causing Doppler effect, diurnal shifts, etc.) result in fluctuations in the carrier frequencies *as received* by the user that may be very much greater than the uncertainty described above.

The broadcasts on 5, 10, and 15 MHz from WWVH are from phased vertical half-wave dipole arrays. They are designed and oriented to radiate a cardioid pattern directing maximum gain in a westerly direction. The 2.5 MHz antenna at WWVH and all antennas at WWV are half-wave dipoles that radiate omnidirectional patterns.

At both WWV and WWVH, double sideband amplitude modulation is employed with 50 percent modulation on the steady tones, 25 percent for the BCD time code, 100 percent for seconds pulses, and 75 percent for voice.

All the WWV and WWVH frequencies are outside the ama-

NATIONAL BUREAU OF STANDARDS
FREQUENCY AND TIME FACILITIES

17-1.

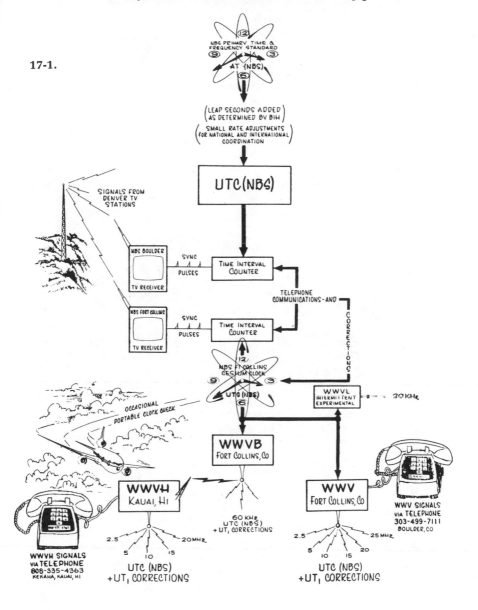

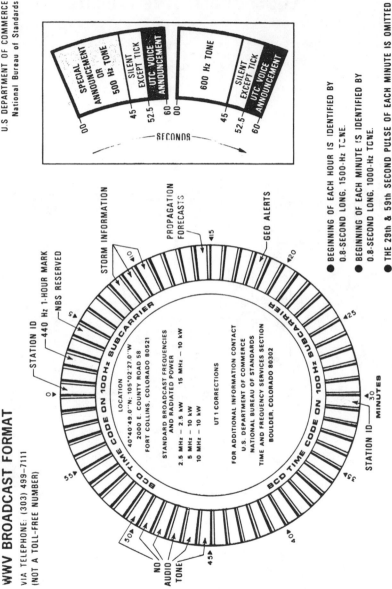

WWV BROADCAST FORMAT

VIA TELEPHONE: (303) 499-7111
(NOT A TOLL-FREE NUMBER)

U.S DEPARTMENT OF COMMERCE
National Bureau of Standards

SPECIAL ANNOUNCEMENT OR 500 Hz TONE

SILENT EXCEPT TICK

UTC VOICE ANNOUNCEMENT

600 Hz TONE

SILENT EXCEPT TICK

UTC VOICE ANNOUNCEMENT

SECONDS

STATION ID

440 Hz 1-HOUR MARK

NBS RESERVED

STORM INFORMATION

PROPAGATION FORECASTS

GEO ALERTS

BCD TIME CODE ON 100Hz SUBCARRIER

LOCATION
40°40'49.0"N: 105°02'27.0"W
2000 E. COUNTY ROAD 58
FORT COLLINS, COLORADO 80521

STANDARD BROADCAST FREQUENCIES
AND RADIATED POWER

2.5 MHz — 2.5 kW   15 MHz — 10 kW
5 MHz — 10 kW
10 MHz — 10 kW

UT 1 CORRECTIONS

FOR ADDITIONAL INFORMATION CONTACT
U.S. DEPARTMENT OF COMMERCE
NATIONAL BUREAU OF STANDARDS
TIME AND FREQUENCY SERVICES SECTION
BOULDER, COLORADO 80302

BCD TIME CODE ON 100 Hz SUBCARRIER

STATION ID

MINUTES

NO AUDIO TONE

• BEGINNING OF EACH HOUR IS IDENTIFIED BY
  0.8-SECOND LONG, 1500-Hz TONE.

• BEGINNING OF EACH MINUTE IS IDENTIFIED BY
  0.8-SECOND LONG, 1000-Hz TONE.

• THE 29th & 59th SECOND PULSE OF EACH MINUTE IS OMITTED

WWVH BROADCAST FORMAT

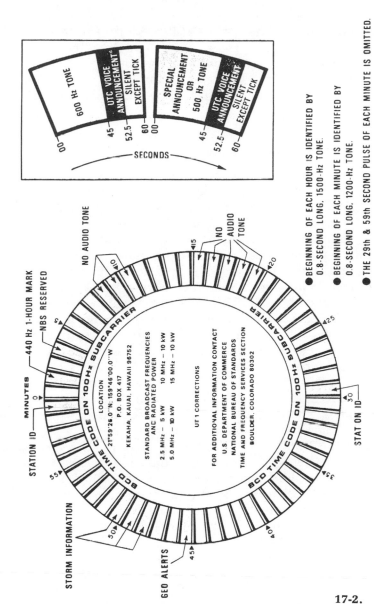

VIA TELEPHONE: (808) 335—4363 (NOT A TOLL-FREE NUMBER)

- BEGINNING OF EACH HOUR IS IDENTIFIED BY 0.8-SECOND LONG, 1500-Hz TONE.
- BEGINNING OF EACH MINUTE IS IDENTIFIED BY 0.8-SECOND LONG, 1200-Hz TONE.
- THE 29th & 59th SECOND PULSE OF EACH MINUTE IS OMITTED.

17-2.

teur bands, but many receivers and transceivers are equipped to receive on either 10 or 15 MHz. All the frequencies are well within the dial spread of general-coverage receivers. An excellent substitute is a small self-contained receiver called the "Timekube" that is pretuned for the 5, 10, and 15 MHz frequencies. This is distributed by Radio Shack stores.

Time Announcements

Voice announcements are made from WWV and WWVH once every minute. To avoid confusion, a man's voice is used on WWV and a woman's voice on WWVH. The WWVH announcement occurs first—at 15 seconds before the minute—while the WWV announcement occurs at 7½ seconds before the minute. Though the announcements occur at different times, the tone markers referred to are transmitted simultaneously from both stations. However, they may not be received at the same time due to propagation effects.

The time referred to in the announcements is "Coordinated Universal Time." It is coordinated through international agreements by the International Time Bureau (BIH) so that time signals broadcast from the many stations such as WWV throughout the world will be in close agreement.

The specific hour and minute mentioned is actually the time at the time zone centered around Greenwich, England, and may be considered generally equivalent to the well-known "Greenwich Mean Time" (GMT). The UTC time announcements are expressed in the 24-hour clock system—i.e., the hours are numbered beginning with 00 hours at midnight through 12 hours at noon to 23 hours, 59 minutes just before the next midnight.

Hams usually keep a 24-hour clock set permanently for UTC because this is the only practical time for arranging and keeping international schedules. West of the Greenwich reference zone, time is earlier; east of it, later. In the U.S.A., the eastern states are five hours behind UTC, the central states six hours, the mountain states seven hours, and the western states eight

17-3. There's no mistaking times in the UTC system if they are spoken correctly. Above: The top clock reads "six hours, twenty three minutes," the bottom one, "thirteen hours, sixteen minutes." Below: Navigator-type wrist watch is also popular with hams. Note that the day starts at the bottom of the dial, "24" being both midnight and the beginning of the new day. The reading here is 1340 UTC.

hours. These times change when daylight saving times are in effect. See Fig. 17-3.

The abbreviation UTC is often shortened to UT, and sometimes to the single letter Z, borrowed from military and commercial practices.

Standard Time Intervals

The most frequent sounds heard on WWV and WWVH are the pulses that mark the seconds of each minute, except for the 29th and 59th seconds pulses which are omitted completely. The first pulse of every *hour* is an 800-millisecond pulse of 1500 Hz. The first pulse of every *minute* is an 800-millisecond pulse of 1000 Hz at WWV and 1200 Hz at WWVH. The remaining seconds pulses are brief audio bursts (5-millisecond pulses of 1000 Hz at WWV and 1200 Hz at WWVH) that resemble the ticking of a clock. All pulses commence at the *beginning* of each second. They are given by means of double-sideband amplitude modulation.

Each seconds pulse is preceded by 10 milliseconds of silence and followed by 25 milliseconds of silence to avoid interference which might make it difficult or impossible to pick out the seconds pulses. This total 40-millisecond protected zone around each seconds pulse is illustrated in Fig. 17-4.

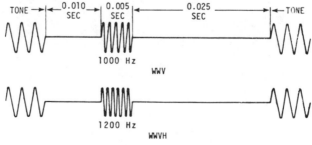

17-4. Pulse separations of time signals from stations WWV and WWVH.

Standard Audio Frequencies

In alternate minutes during most of each hour, 500 or 600 Hz audio tones are broadcast. A 440 Hz tone, the musical note A above middle C, is broadcast once each hour. In addition to being a musical standard, the 440 Hz tone can be used to provide an hourly marker for chart recorders or other automated devices.

Forty-five-second announcement segments are available on a subscription basis to other federal agencies to disseminate official and public service information. The accuracy and content of these announcements are the responsibility of the originating agency, not necessarily NBS.

Marine Storm Warnings

Weather information about major storms in the Atlantic and eastern North Pacific are broadcast in voice from WWV at 8, 9, and 10 minutes after each hour. Similar storm warnings covering the eastern and central North Pacific are given from WWVH at 48, 49, and 50 minutes after each hour. An additional segment (at 11 minutes after the hour on WWV and at 51 minutes on WWVH) may be used when there are unusually widespread storm conditions. The brief messages are designed to tell mariners of storm threats in their areas. If there are no warnings in the designated areas, the broadcasts will so indicate. The ocean areas involved are those for which the U.S. has warning responsibility under international agreement. The regular times of issue by the National Weather Service are 0500, 1100, 1700, and 2300 UTC for WWV and 0000, 0600, 1200, and 1800 UTC for WWVH. These broadcasts are updated effective with the next scheduled announcement following the time of issue.

Mariners might expect to receive a broadcast similar to the following:

North Atlantic weather west of 35 West at 1700 UTC: Hurricane Donna, intensifying, 24 North, 60 West, moving northwest, 20 knots, winds 75 knots; storm, 65 North, 35 West, moving east, 10 knots; winds 50 knots, seas 15 feet.

"Silent" Periods

These are periods with no tone modulation. However, the carrier frequency, seconds pulses, time announcements, and 100-Hz BCD time code continue. The main silent periods extend from 45 to 51 minutes after the hour on WWV and from 15 to 20 minutes after the hour on WWVH. An additional 3-minute period from 8 to 11 minutes after the hour is silent on WWVH.

BCD Time Code

A binary coded decimal (BCD) time code is transmitted continuously by WWV and WWVH on a 100-Hz subcarrier. The 100-Hz subcarrier is synchronous with the code pulses so that 10-millisecond resolution is attained. The time code provides a standard timing base for scientific observations made simultaneously at different locations. It has application, for example, where signals telemetered from a satellite are recorded along with the time code pulses. Data analysis is then aided by having accurate, unambiguous time markers superimposed directly on the recording.

The WWV/WWVH time code format presents UTC information in serial fashion at a rate of one pulse per second. Groups of pulses can be decoded to ascertain the current minute, hour, and day of year. While the 100-Hz subcarrier is not considered one of the standard audio frequencies, the code does contain the 100-Hz frequency and may be used as a standard with the same accuracy as the audio frequencies.

UT1 Time Corrections

The UTC time scale broadcast by WWV and WWVH runs at a rate that is almost perfectly constant because it is based on ultrastable atomic clocks. This time scale meets the needs of most users. Somewhat surprisingly, however, some users of time signals need time which is not this stable. In applications such as very precise navigation and satellite tracking, which must be referenced to the rotating earth, a time scale that speeds up and slows down with the earth's rotation rate must be used. The particular time scale needed is known as UT1 and is inferred from astronomical observations.

To be responsive to these users, information needed to obtain UT1 time is included in the UTC broadcasts. This occurs at two different levels of accuracy. First, for those users needing to know UT1 only to within about one second (this includes nearly all boaters/navigators), occasional corrections of exactly one second—called "leap" seconds—are inserted into the UTC time scale whenever needed to keep the UTC time signals within ±0.9 second of UT1 at all times. These leap seconds may be either positive or negative and are coordinated under international agreement by the International Time Bureau (BIH) in Paris. Ordinarily, a positive leap second must be added about once per year (usually on June 30 or December 31), depending on how the earth's rotation rate is behaving in each particular year.

The second level of correction is included in the UTC boradcasts for the very small number of users who need UT1 time to better than one second. These corrections, in units of 0.1 second, are encoded into the broadcasts by using double ticks or pulses after the start of each minute. The amount of correction is determined by counting the number of successive double ticks heard each minute. The 1st through the 8th seconds ticks indicate a "plus" correction, and the 9th through the 16th, a "minus" correction. For example, if the 1st, 2nd, and 3rd ticks are doubled, the correction is "plus" 0.3 second: UT1 = UTC + 0.3 second, or if UTC is 8:45:17, then UT1 is 8:45:17.3. If the 9th, 10th, 11th, and 12th ticks are doubled, the correction is "minus" 0.4 second, or as in the above example, UT1 = 8:45:16.6.

Station WWVB

WWVB transmits continuously on a standard radio carrier frequency of 60 kHz. Standard time signals, time intervals, and UT1 corrections are provided by means of a BCD time code. The station is located on the same site as WWV. Effective coverage area is the continental U.S.

The frequency of WWVB is normally within its prescribed value to better than 1 part in 100 billion. Deviations from day to day are less than 5 parts in 1,000 billion. Effects of the propagation medium on received signals are relatively minor at low frequencies; therefore, frequency comparisons to better than 1 part in 100 billion are possible using appropriate receiving and averaging techniques.

WWVB identifies itself by advancing its carrier phase 45 degrees at 10 minutes after every hour and returning to normal phase at 15 minutes after the hour. WWVB can also be identified by its unique time code.

The effective radiated power from WWVB is 13 kw. The antenna is a 122-meter, top-loaded vertical installed over a radial ground screen.

Time by Telephone

A little-known aspect of the NBS services is that the WWV and WWVH broadcasts can be heard via telephone. By calling 303-499-7111 in Boulder or 808-335-4363 in Hawaii you can hear three minutes of the audio portions of the transmissions. Depending on where you live, calls can cost as little as 30 cents for three minutes to about 75 cents for the longest possible distance.

Canada's Time Service

The National Research Council of Canada radio time signals broadcast from Ottawa over station CHU have the following characteristics: Continuous transmission is made on three frequencies, 3330, 7335 and 14,670 kHz. Transmitter power output is 3 kw on 3330 and 14,670 kHz, and 10 kw on 7335 kHz. Vertical antenna systems are employed.

The three transmitter frequencies and time signals are derived from a cesium frequency standard which is referred daily to the Canadian cesium primary standard.

Take particular note of the 7335-kilohertz frequency, because this is just within the tuning range of the 40-meter band on many ham-band transceivers and receivers. The upper limit of the band is 7300 kHz, so CHU is only a few dial divisions further up. See Fig. 17.5.

CHU DATA TRANSMISSION SEQUENCE

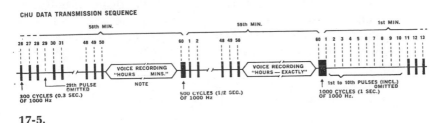

17-5.

The seconds pulses consist of 300 cycles of a 1000 Hz tone. The beginning of the pulse marks the exact second. The zero pulse of each minute is ½ second long, and the zero pulse of the hour is one second long. The pulses occur at the rate of one each second with the following exceptions:—

1. The 29th pulse of each minute is omitted.
2. The 51st to 59th pulses inclusive of each minute are omitted. During this interval station identification and time is announced by voice.
3. The 1st to 10th pulses inclusive are omitted on the first minute of each hour.

A voice recording of the time occurs each minute in the ten-second gap between the 50th and 60th second. It refers to the beginning of the minute or hour pulse that follows. The announcement is on the 24-hour system, alternating in French and English.

CHU is operated by the National Research Council of Canada, Ottawa, Ontario, Canada, K1A-OS1. The station is located on the outskirts of Ottawa at 45°17'47" North, 75°45'22" West.

Measuring Distance

A brightly colored map of the world measuring 2 x 3 or 3 x 4 feet makes attractive wallpaper in a radio shack, but it is not reliable for measuring distance. No flat map can be, because the earth is not flat. If you want to know the distance between you and those chaps in South America, Europe, and Asia whom you worked on 20 meters, you need a map shaped like the earth. Not surprisingly, such a map is called a *globe*. See Fig. 17-6.

If your shack is of average size, a 12-inch globe is about right. Other sizes up to about three feet are available.

17-6. How far is it from here to there? Lay a string between the two points on a globe, check this against the scale on the latter, and you have the answer.

18

Mobile Operation

Amateur communication is not confined to fixed stations in homes or apartments. You can "go mobile" in any vehicle of your choice: bicycle, motorcycle, automobile, camper, boat, balloon, or airplane! Your regular license covers any such station in motion; you simply add the word *mobile* after your call letters. Operation is almost entirely on voice, but there is nothing to prevent you from using CW.

Most mobile rigs are, of course, in cars. The big advantage they have over fixed installations is that every new hilltop or open stretch of road presents a new set of operating conditions and the opportunity for new and unusual local and DX contacts. For many hams, especially those who live in apartment houses and have an insoluble antenna problem, mobile is their complete salvation.

Mobile stations are limited in physical size by the dashboard and the seating arrangment of the car, in power rating by the capacity of the battery and its charging generator, and in radiation efficiency by the enforced use of small, short antennas. In spite of all this, remarkable work can be done on various bands. It is often freakish, but never uninteresting! The editor of this book once stopped his car in Times Square, New York City, completely hemmed in by tall buildings and blinking electric signs. In less than an hour, using a 50-watt transmitter and an 8-foot whip antenna, he worked stations in England, Germany, North Africa, and the Canal Zone! On another occasion, while parked on a Denver hill a full mile above sea level, with a clear view in every direction to the ho-

18-1. This large Collins KWM-2 transceiver, of the conventional tube type, mounts to a special plug-in base fastened to the floor of the car. Its current drain on the car's battery is considerable.

18-2. By contrast, this Atlas-210 all solid-state five-band transceiver is not much larger than a cigar box, and needs little current.

rizon, he shouted himself hoarse without getting a single reply during more than an hour of dial twiddling. Flying over Ohio in a friend's airplane, he once carried on a conversation with a lad on a freighter near the coast of Japan, and this continued right up to the moment the plane touched down in Washington, D.C.

Equipment for Mobile

For years most mobile ham sets for the normal 10- through 80-meter bands were merely modified versions of fixed-station AM equipment of the tube type. They worked well, but their power packs took large amounts of current from the vehicle's electrical system; much more, in fact, than many storage batteries and their related charging systems could furnish without burning up. It was also common to use separate receivers and transmitters. In many cases the receive unit was actually a small shortwave converter that fed into the car's broadcast receiver, from which the ham signals would issue.

All this changed dramatically with the solid-state revolution. Mobile sets for the 10-80 bands not only went exclusively SSB but also shrank to little more than the size of cigar boxes, with greatly reduced current requirements. All are now transceivers. Some are small enough to fit in the glove compartments of full-size cars, but more commonly they are placed under the dash on pivoted mounts within easy reach of both the driver and any passenger alongside.

The mounts are often of the plug-in type and permit the entire transceiver to be pulled out quickly for concealment in the trunk of the car, and replaced just as quickly. This is a worthwhile security measure and reduces the chance of loss to roving thieves.

All mobile rigs have built-in loudspeakers and plug-in hand microphones for easy push-to-talk operation. They need only two connections, one to the coax from the antenna and the other to the "hot" (ungrounded) side of the car's battery.

Very conveniently, the latter connection can be made with a short cord to the nearby cigarette lighter on the dash.

Take Care of the Battery!

Mobile rigs are great for field-day exercises and similar outdoor activities and also for emergency communications during public disasters such as fires, storms, power failures, etc. However, it is important to remember that most car batteries are designed only to provide short but heavy blasts of current just to get the engines started; then the generators take over and start to recharge them. If you've ever left the lights of a car on overnight, you know what a problem you have in the morning!

For serious field operation many hams treat themselves to a small gasoline-fueled generator and use this independently of the car's electrical system. Units of this type weigh as little as sixteen pounds and therefore are easily stowed in the trunk. They are popular with owners of RV's (recreational vehicles) and are available from most RV dealers and some radio dealers.

Road Safety Precautions

Certain important precautions must be observed in mobile operation. First, from the safety standpoint it is highly advisable to refrain from using the rig at all in heavy traffic. If an interesting contact develops unexpectedly, the smart move is to get off the road and to operate with the car stationary. Let the engine idle during reception and race it slightly during transmission to bring the charging generator on. In cold weather especially, this might save you the embarrassment of a stalled car at the end of a long QSO.

In a true transceiver, the operating frequency is precisely the same for both transmission and reception at any one dial

setting; the operator cannot send on one frequency and receive on another. This seeming inflexibility is an operating
advantage rather than a shortcoming, because probably 90
percent of all ham contacts are made on a common frequency
rather than on two different settings. The latter mode of
operating is known as *working cross-band* and has a place in
fixed-station rather than mobile practice.

Two Stations in One

With the reduction in the size of the mobile gear for both the
10-80 meter SSB bands and the 2-meter FM band, it is now
feasible to fit an ordinary passenger car with two independent
transceivers, and to enjoy their separate facilities as circumstances permit. The low-frequency whip can remain on the
rear bumper, while the much smaller FM rod can be mounted
either on the rain gutter over a door or in the hole in a front
fender usually reserved for broadcast reception. For the latter
purpose, many late-model cars use a thin wire imbedded in
the front windshield, so the space on the fender is there to
use.

The ideal shack-on-wheels is, of course, any recreational
vehicle. A layout that has become popular consists of a 2-
meter rig paired with a CB unit, within arm's reach of the
driver, and a low-frequency transceiver in the eating area, for
use only during stops. And there's lots of flat roof area for a
small antenna farm.

Whip Antennas

For 10 meters and lower, the mobile antenna is usually a
slender metal whip, one-quarter wavelength long, with the
body of the car acting as the other quarter-wave. This operation is similar to that of a fixed vertical antenna working
against the ground. An 8-foot whip is about right for 10

18-3. Separate receiver and transmitter are used in this mobile installation. Part of glove compartment has been sacrificed to make room for the former. In this position it is within reach of operator.

18-4. Bil Harrison, W2AVA, of New York, doesn't take his hands off the wheel when he's out mobiling. He wears a neck mike and has an earplug receiver, and uses his left foot to operate a switch that controls his transceiver. Latter is mounted on floor behind him.

13-4. Bil Harrison, W2AVA, of New York, doesn't take his hands off the wheel when he's out mobiling. He wears a neck mike and has an earplug receiver, and uses his left foot to operate a switch that controls his transceiver. Latter is mounted on floor behind him.

18-5. A telephone operator's chest-type microphone lends itself to mobile use because it remains in correct speaking distance from the driver's mouth. He presses switch on the mike's base to actuate the transmitter under the dash. Receiver is farther to left.

meters and doesn't protrude too far above the top of the car. Above 10 meters, however, a simple whip becomes impractically long. The practice is to use a relatively short whip or rod—as short as 4 feet—and to add loading or trap coils to it to make it work either on one band or on several. Coax cable is used exclusively in mobile installations to join the antenna to the set proper.

Mobile antennas are supported in short, husky insulators that are mounted either on the rear bumper, by means of suitable clamps, or in a hole in a rear fender or the trunk lid. Their characteristics are affected by their position in relation to the body. Fortunately, since they are readily accessible, it is easy to experiment with their length and their loading coils.

18-6. Below: What better place for a mobile station than a mobile home? Use battery power on the road; then plug in to the AC line when you stop for the night in a mobile-home park. In this typical set-up, a whip antenna rides high and clear without obstructing the driver's vision. It can be folded when the vehicle must pass through a covered toll booth or under a bridge.

Call-Letter License Plates

All but a few states permit hams to obtain automobile license plates incorporating their FCC call letters. These usually cost a little more than regular plates, but of course they are highly distinctive and much prized. An actual rig in the car is not a prerequisite, but a copy of your FCC ticket may be. Check with your motor vehicle department.

18-7. As Irving Strauber has learned, there's nothing like your call letters on your license plate to attract the attention of other hams for "eyeball QSO's," that is, conversations in person.

19

Citizens Band Radio

In 1958 the FCC opened the frequencies between 26.96 and 27.26 megahertz for low-power voice operation on 23 specific channels within these limits.* Called the "Citizens Band," this 300-kilohertz slice of the radio spectrum was intended primarily for short-range communication between vehicles of any sort and related base stations in the offices of small businesses or in private homes. Also authorized was the use of small hand-held "walkie-talkies" for person-to-person, person-to-vehicle, and person-to-base contacts.

CB was never intended to replace or supplement any of the busy ham bands, and in fact FCC regulations specifically prohibited hobby-type conversations. However, because anybody could obtain a valid CB license without undergoing a technical or operating examination of any kind, the legitimate users of CB were plagued from the start by interference from thoughtless gabbers.

Acknowledging the impossibility of monitoring several million stations, in 1975 the FCC authorized hobby-CB by designating Channel 11 as the exclusive calling frequency for CB enthusiasts. Once they establish contact, they must switch quickly to any one of the other channels (except No. 9) that is not busy at the moment. With this arrangement, hobby-operating discipline has improved vastly.

The no-code, test-free CB ticket is only one of the differences between CB and ham radio. CBers are limited to

---

* The band was later widened to 40 channels.

transmitter power of five watts and to a single band; hams may use as much as 1,000 watts on a dozen or more bands.

More significant facts. A ham license does not give the holder CB privileges. A CB license does not count in any way as credit toward a ham ticket. However, a person can hold *both* licenses and use them independently with the appropriate equipment.

Why is Channel 9 exempt from hobby use? Because it is the national emergency frequency, and is monitored around the clock by CB clubs and also by many public safety departments such as police and fire and by independent civilians engaged in vehicular operations of many kinds. Since 1962 there has been a voluntary organization of CBers called REACT (for "Radio Emergency Associated Communication Teams"), which performs extraordinary free services for the traveling public. In a period of thirteen years the teams handled almost

19-1. Out of gas, two tires flat, or radiator leaking? With a CB transceiver in the car, this driver is fairly sure of getting help if he calls on emergency Channel 9 and gives his location. Note thin whip antenna on top of car. (*Radio Shack*)

12 million calls relating to highway accidents alone, and also assisted in many relief efforts during floods, hurricanes, earthquakes, and other disasters. Channel 9 by itself more than justifies the existence of CB. Any calls you hear on it have top priority.

In 1975 REACT was reorganized as a nonprofit corporation called REACT International, Inc. It receives support from many safety-minded people and companies, including, not surprisingly, automobile manufacturers.

While CB as a hobby offers only limited possibilities and cannot be compared fairly with ham radio, as a security measure for mobile-minded Americans it is literally a lifesaver. To

19-2. The immobilized driver is in luck, because the enterprising owner of a nearby service station monitors Channel 9 on his Realistic Model TRC-55 base station, made by Radio Shack, and hears the call for assistance.

call for help from a stalled car, to report an overturned truck, to warn about a blocked road, to ask for directions to a motel, a drugstore, a hospital, etc., you only have to turn the channel selector of a CB transceiver to 9 and talk into the mike. Anybody who owns a TV set can be taught in five minutes how to use CB. Use it just once in an emergency and it will have paid for itself many times over.

Mobile CB is not limited to cars and trucks. It is also valuable on pleasure boats, especially those used on inland waters where there are no Coast Guard facilities. It supplements but does not replace regular marine radio services on special marine frequencies.

19-3. Lost on a strange road? Since mother knows how to turn on and operate the CB rig her husband installed in the car for her safety, she picks up the mike, gives her call sign and location, and explains her problem. (*Radio Shack*)

Being limited to five watts and operating with AM (amplitude modulation), conventional CB transceivers are generally smaller and much less expensive than ham transceivers. Current drain on a vehicle's battery averages about two amperes. Transmission starts with a small hand-held mike, and the received signals come out of a loudspeaker in the lower part of the transceiver case. More advanced rigs use a standard telephone-type handset, with a push-to-talk switch in the handle.

CB is not restricted to amplitude-modulated equipment. In fact, the FCC encourages the use of SSB (single sideband) because it makes possible the double-sharing of any frequency without mutual interference. One two-way conversation can be held on upper sideband (USB) while another is going on lower sideband (LSB). This arrangement requires that all four stations be of the SSB type; an intermix of SSB and older straight AM sets is not practical. There is a strong likelihood that some existing CB channels, and certainly many new ones as they become available, will be reserved exclusively for SSB, so that its full advantages can be obtained.

19-4. An alert volunteer CB operator at a REACT station, surrounded by maps, pinpoints the position of the stranded mother and gives her the road information or other assistance that she needs.

It is easy to spot a CB-equipped car on the road by its antenna, usually mounted on the back edge of the trunk lid. This is either a slim whip about four feet long or a stubby rod about two feet long with a noticeable bulge in the center; the bulge represents a loading coil.

How far can you talk with CB? This depends on local conditions. Between cars on flat terrain, with no large buildings or hills in the way, figure on 4 to 8 miles. On water you can do better. Base-to-mobile distances depend mostly on the height of the base station's antenna. We are considering only really reliable communication, not just occasional contacts, by the part of the transmitted radio waves that hugs the ground. On the CB frequencies, as on most of the ham bands, another part of the waves takes off from the antenna at an angle toward the sky, skips merrily through the upper atmosphere, and often bounces back to earth 1,000, 2,000, or 4,000 miles away. You are quite likely to hear and even talk to stations at these astonishing distances. Their signals might be rip-roaring loud for five or fifteen minutes and then disappear completely as

19-5. Not much larger than a telephone book, the Lafayette Telsat 1023 CB transceiver on the desk of this businessman enables him to keep in close touch with a fleet of service vans.

the angle of reflection from the sky changes.

The range of walkie-talkies is perhaps a mile or two under good conditions, meaning fresh batteries and a clear point-to-point path.

Incidentally, no license at all is required for walkie-talkies with a power rating of 100 milliwatts (one-tenth of one watt) or less, but they are hardly more than expensive toys.

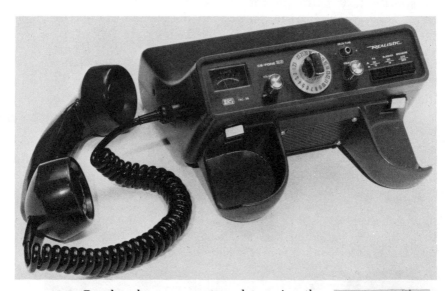

19-6. People who are accustomed to using the telephone like this type of mobile CB transceiver because the handset fits naturally in their palms. A built-in loudspeaker can be switched on to let other passengers in the car hear conversations. The Radio Shack CB-Fone, Model TRC-56.

19-7. Hand-held "walkie-talkies" are generally of this vertical style, with an antenna that pulls out from the top of the case. Microphone is behind grille at top; loudspeaker at bottom. The Realistic TRC-27.

The FCC call signs for CB stations consist of three letters and three or four numbers. They are issued strictly in sequence, and special combinations cannot be ordered.

FCC application blanks for CB licenses are available free of charge from virtually all stores that sell CB equipment. If you think you might want a car rig for your next trip cross-country, get your form off to the FCC right away, as there will undoubtedly be a delay of several months before it is processed. There are more CB stations than all other kinds combined, and then some, and the FCC is hard-pressed to keep up with the paperwork.

19-8. CB is especially effective for owners of either large or small boats because transmission over water, in the absence of intervening buildings, is highly reliable even with low-power equipment.

CB "Ten" Code

Adapted from a brevity code long used by police, fire, and other public-service agencies, this consists simply of combinations of digits, all starting with 10, that have the arbitrary meanings listed below. The correct use of this CB shorthand greatly speeds up communication, especially during road emergencies. Two of the signals, 10-18 and 10-25, are questions only. The others can be statements, requests, or questions, depending on the voice inflections of the operators.

The Ten code is completely unofficial; that is, it is not required for CBers by the FCC as the International Morse Code is for amateurs. In fact, hams who also have CB rigs tend to lapse into ham lingo when they talk to each other and they often mix the Q and Ten codes, to the consternation of many listeners.

10-1	Receiving poorly	10-37	Wrecker needed at _____
10-2	Receiving well	10-38	Ambulance needed at _____
10-3	Stop transmitting	10-39	Your message delivered
10-4	OK, message received	10-41	Please tune to channel _____
10-5	Relay message	10-42	Traffic accident at _____
10-6	Busy, stand by	10-43	Traffic tieup at _____
10-7	Out of service, leaving air	10-44	I have a message for you (or _____)
10-8	In service, subject to call	10-45	All units within range please report
10-9	Repeat message	10-50	Break channel
10-10	Transmission completed, standing by	10-60	What is next message number
10-11	Talking too rapidly	10-62	Unable to copy, use phone
10-12	Visitors present	10-63	Net directed to _____
10-13	Advise weather/road conditions	10-64	Net clear
10-16	Make pickup at _____	10-65	Awaiting your next message/assignment
10-17	Urgent business	10-67	All units comply
10-18	Anything for us?	10-70	Fire at _____
10-19	Nothing for you, return to base	10-71	Proceed with transmission in sequence
10-20	My location is _____	10-73	Speed trap at _____
10-21	Call by telephone	10-75	You are causing interference
10-22	Report in person to _____	10-77	Negative contact
10-23	Stand by	10-81	Reserve hotel room for _____
10-24	Completed last assignment	10-82	Reserve room for _____
10-25	Can you contact _____	10-84	My telephone number is _____
10-26	Disregard last information	10-85	My address is _____
10-27	I am moving to channel _____	10-89	Radio repairman needed at _____
10-28	Identify your station	10-90	I have TVI
10-29	Time is up for contact	10-91	Talk closer to mike
10-30	Does not conform to FCC rules	10-92	Your transmitter is out of adjustment
10-32	I will give you a radio check	10-93	Check my frequency on this channel
10-33	Emergency traffic at this station	10-94	Please give me a long count
10-34	Trouble at this station, help needed	10-95	Transmit dead carrier for 5 sec.
10-35	Confidential information	10-99	Mission completed, all units secure
10-36	Correct time is _____	10-200	Police needed at _____

CB Slang

Most of the colorful expressions that make the CB bands so interesting originated with truckers, who found their microphones a quick antidote for highway boredom. Some of their terms have even found their way into the common language; for example, *Smokey the Bear* for state police, *hammer* for gas pedal, *ratchet jaw* for an overly talkative operator, etc. New words and phrases are heard almost daily.

The following list does not claim to be complete, but at least it will enable a new listener to make some sense out what he hears.

Advertising—Police car with its lights on.

Back Door—Rear vehicle of two or more running together (via CB).

Beat the Bushes—Vehicle driving ahead of a group and going just enough over the speed limit (but not fast enough to get a ticket) to bring out any hidden police cars to investigate. Lead vehicle watching for speed traps.

Bear—A police officer. See "Smokey."

Bear Cave—Also Bear Den. Any police station.

Bear in the Air—Police aircraft used to clock highway traffic.

Bean Store—Restaurant or road stop where food is served.

Big 10-4—Very much in agreement; "You said a mouthful!"

Bottle Popper—Beverage (usually beer) truck.

Boulevard—Highway.

Break—Request to use channel, often given with channel number, i.e. "Break channel one-four" (I'd like to make a call on channel 14).

Breaker—Station requesting a break.

Brown Bottles—Beer.

Brush Your Teeth and Comb Your Hair—Radar unit ahead.

Bushel—1,000 pounds.

Camera—Police radar unit.

Charlie—The FCC. Also, "Uncle Charlie."

Chicken Coop—Truck weighing station.

Chicken Inspector—Weight station inspector.

Clean—No police in sight.
Clear—Communications completed.
Cotton Picker—Used in place of any stronger terms. i.e. "That cotton picker just cut me off!"
County Mounty—County police or sheriff.
Cowboy Cadillac—An El Camino or Ford Ranchero.
Cut Some Z's—Get some sleep.

Drop the Hammer—Accelerate.

Ears—A CB radio or the antenna for a CB radio.
Eighteen Wheeler—Any tractor-trailer truck regardless of actual number of wheels.
Eye-in-the-Sky—Police aircraft.
Evel Knievel—Motorcycle rider.

Fat Load—Overweight load.
Feed the Bears—Pay a traffic ticket.
Fifty Dollar Lane—Leftmost or passing lane.
Flip-Flop—Return trip, or "U" turn.
Fluff-Stuff—Snow.
Fly in the Sky—Police aircraft.
Four Wheeler—Any passenger vehicle with four wheels.
Front Door—Lead vehicle of two or more running together (via CB).

Good Numbers—As in "All the good numbers to you." Best regards and good wishes.
Got Your Ears On?—Are you listening to your CB radio?
Grass—Median strip or alongside of road.
Green Stamps—Money.
Green Stamp Road—Toll road.
Ground Clouds—Fog.

Haircut Palace—Bridge or overpass with low clearance.
Hammer—Accelerator pedal.
Handle—Name used on CB radio.
Harvey Wallbanger—Reckless driver.
Holding on to Your Mud Flaps—Driving right behind you.

Hole in the Wall—Tunnel.
Home 20—Where you live. Home town.

Invitations—Police traffic citations; tickets.

Keep the Shiny Side Up and the Dirty Side Down—Don't have an accident.

Land Line—Telephone.
Local Yocal—City police.
Loose Board Walk—Bumpy road.

Mama Bear—Policewoman.
Mercy!—Expletive exclamation.
Mile Marker—Milepost on interstate highways.
Mix-Master—Highway cloverleaf.
Modulate—Talk.
Monfort Lane—Passing lane.

Nap Trap—Rest area or motel.
Negatory—Negative.

On the Side—Standing by and listening.

Pickum-Up—Pickup truck.
Picture Taker—Police radar.
Plain Wrapper—Unmarked police car. Usually given as: Smokey in a plain brown wrapper (brown car), plain green wrapper (green car), etc.
Portable Chicken Coop—Portable truck scale.
Portable Parking Lot—Auto carrier.
Post—Milepost on interstate highways.
Pregnant Roller Skate—Volkswagen.
Pull the Big Switch—To turn off the CB radio.

Radio—A CB transceiver.
Ratchet Jaw—Overly talkative CBer.
Rig—CB radio; also truck tractor.
Rocking Chair—Vehicle between lead "front door" and rearmost "back door" vehicles.
Roger Ramjet—Driver of a car going well over the speed limit.

Rollerskate—Small car such as a compact or import.
Rolling Road Block—Vehicle going under the speed limit and holding up traffic.

Sailboat Fuel—Running empty.
Seatcovers—Passengers.
Shake the Trees and Rake the Leaves—Lead vehicle watch ahead, rear vehicle watch behind.
Skating Rink—Slippery road.
Smokey—Any police officer.
Smokey the Bear—State police.
Smokey's Got Ears—Police with CB radio.
Spy in the Sky—Police aircraft.
Super Skate—High performance car, Corvette or other sports car.
Super Slab—Major highway.

Taking Pictures—Police using radar.
Ten-four—Affirmative (see ten-code, page 291)
Threes—Best regards.
Threes and Eights—Best regards, love and kisses (those "good numbers").
Tijuana Taxi—Police car with lights and identification on it.
Town—Any city, regardless of size, i.e. New York town, Dallas town, Podunk town, etc.
Truck 'em Easy—Have a good trip.
Twenty—Location (10-20).
Twisted Pair—Telephone.
Twister—Highway interchange.
Two Wheeler—Motorcycle.

Uncle Charlie—The FCC.

Wall-to-Wall Bears—Heavy police patrol.
Wall-to-Wall and Treetop Tall—Receiving you loud and clear.
Willy Weaver—Drunk driver.
Window Washer—Rainstorm.

X-ray Machine—Police radar.

Appendix A
Insurance Requirements

There is much confusion over the "National Electric Code." This is not an official United States Government document, as many people believe, but a publication of the National Fire Protective Association. The latter is a consortium of insurance companies, appliance manufacturers, and builders of homes and other structures who are interested in keeping you alive and your property intact so that they won't have to pay out large sums for losses covered by policies.

The recommendations made by the Code, and that's all they are, are quite sensible and represent safety measures that any person in his right mind would take anyway. However, the extent to which they affect any insurance policy depends on its precise wording and on the enforcement by municipal authorities of any pertinent Code recommendations that have been incorporated into local building regulations. To be safe, check with your insurance agent.

The full Code contains more than 500 pages, much of it of no interest to individuals. However, a worthwhile purchase for $2 is an abridged edition entitled "One- and Two-Family Residential Electrical Code," available from the National Fire Protective Assn., 60 Batterymarch Street, Boston, Mass., 02110.

Appendix B
Frequently Used Abbreviations of Basic Units and Terms

There is no national or international standardization of radio terms, symbols, and abbreviations. The practices of even the most advanced engineering societies and publications are often inconsistent in themselves. A single letter might be used to represent several entirely different terms, and it might appear in some places as a capital and in others as a small letter. In many cases it is necessary to study the circuit circumstances and to relate the abbreviations to them. For example, if a power transformer or a motor is under discussion and the abbreviation *PF* or *pf* appears, it is safe to assume that it means *power factor*. If the same abbreviation is connected with capacitors, it undoubtedly means *picofarad*.

Some abbreviations are unmistakable in either capitalized or lower-case form. For example, DC and dc, or AC and ac, are easily recognized as *direct current* and *alternating current*, respectively. However, m, which usually stands for *milli*, sometimes is also used for *micro* if the typesetter does not have Greek letters in his shop. Thus, *ma* is likely to represent either *milliampere* or *microampere*.

The capital letter M in formulas means *mutual inductance*, but as a unit prefix it is interpreted as *mega*. Thus, MHz means one million hertz.

The most commonly used terms and their abbreviations are given in the following list. Note that periods are entirely absent.

Alternating current	AC	Antenna	ant
American Wire Gauge	AWG	Audio frequency	AF
Ampere	amp	Automatic frequency	
Amplification factor	μ (mu)	control	AFC
Amplitude modulation	AM	Automatic load control	ALC

Automatic noise limiter	ANL	Micromicrofarad	$\mu\mu$f
Automatic volume control	AVC	Microvolt	μv
		Microwatt	μw
Beat-frequency oscillator	BFO	Milliampere	ma
		Millihenry	mh
Brown and Sharp wire gauge (now American Wire Gauge)	B&S	Modulated continuous wave	MCW
Capacitance	C	Mutual inductance	M
Capacitive reactance	X_C	Ohm	Ω (omega)
Cathode-ray tube	CRT	Phase displacement (degrees)	θ (theta)
Centimeter	cm	Picofarad	pf
Continuous wave	CW	Power	P
Current	I	Power amplifier	PA
Decibel	db	Power factor	PF
Direct current	DC	Push-to-talk control	PTT
Double-pole, double-throw	DPDT	Radio frequency	RF
		Reactance	X
Double-pole, single-throw	DPST	Resistance	R
		Root-mean-square	rms
Double-sideband suppressed carrier	DSB	Single-pole, double-throw	SPDT
Electromotive force	emf	Single-pole, single-throw	SPST
Farad or frequency	f		
Frequency modulation	FM	Single-sideband suppressed carrier	SSB
Ground	gnd	Standing-wave ratio	SWR
Henry	h	Tuned radio frequency	TRF
Hertz	Hz	Ultrahigh frequency	uhf
High frequency	hf	Upper sideband	USB
Impedance	Z	Vacuum-tube voltmeter	VTVM
Inductance	L	Variable-frequency oscillator	VFO
Inductive reactance	X_L		
Intermediate frequency	IF	Very high frequency	vhf
Kilohertz	kHz	Voice-operated transmission	VOX
Kilovolt	kv		
Kilowatt	kw	Volt	v
Load resistance	R_L	Voltage	E
Lower frequency	lf	Volt-ohmmeter	VOM
Lower sideband	LSB	Watt	w
Medium frequency	mf	Wavelength	λ (lambda)
Megahertz	MHz		
Megohm	$M\Omega$		
Meter	m		
Microampere	μa		
Microfarad	μf		
Microhenry	μh		

The Greek Alphabet

Practically all modern ham transmitters contain an important circuit element called a *pi network*. It has this name because the arrangement of its components resembles the shape of the Greek letter *pi* (pronounced *pie*). Many other letters of this ancient alphabet have applications in technology, so you should learn to recognize them. Here they are:

Name	Capital	Lower Case	Represents
Alpha	A	α	Angles. Area. Coefficients.
Beta	B	β	Angles. Flux density. Coefficients.
Gamma	Γ	γ	Conductivity. Specific gravity.
Delta	Δ	δ	Variation. Density.
Epsilon	E	ϵ	Base of natural logarithms.
Zeta	Z	ζ	Impedance. Coefficients. Coordinates.
Eta	H	η	Hysteresis coefficient. Efficiency.
Theta	Θ	θ	Temperature. Phase angle.
Iota	I	ι	Unit vector.
Kappa	K	κ	Dielectric constant.
Lambda	Λ	λ	Wavelength in meters.
Mu	M	μ	Micro. Amplification factor. Permeability.
Nu	N	ν	Reluctivity.
Xi	Ξ	ξ	Coordinates.
Omicron	O	o	. . . .
Pi	Π	π	3.1416 (Ratio of circumference to diameter).
Rho	P	ρ	Resistivity.
Sigma	Σ	σ	Sign of summation.
Tau	T	τ	Time constant. Time phase displacement.
Upsilon	Υ	υ	. . . .
Phi	Φ	φ	Magnetic flux. Angles.
Chi	X	χ	Electric susceptibility. Angles.
Psi	Ψ	ψ	Dielectric flux. Phase difference.
Omega	Ω	ω	Capital, ohms. Lower case, angular velocity.

Appendix C

Television Channel Frequencies

P / S (MHz)	Channel No.	Freq. Limits
P 55.25 / S 59.75	2	54–60
P 61.25 / S 65.75	3	60–66
P 67.25 / S 71.75	4	66–72
P 77.25 / S 81.75	5	76–82
P 83.25 / S 87.75	6	82–88
P 175.25 / S 179.75	7	174–180
P 181.25 / S 185.75	8	180–186
P 187.25 / S 191.75	9	186–192
P 193.25 / S 197.75	10	192–198
P 199.25 / S 203.75	11	198–204
P 205.25 / S 209.75	12	204–210
P 211.25 / S 215.75	13	210–216
P 471.25 / S 475.75	14	470–476
P 477.25 / S 481.75	15	476–482
P 483.25 / S 487.75	16	482–488
P 489.25 / S 493.75	17	488–494
P 495.25 / S 499.75	18	494–500
P 501.25 / S 505.75	19	500–506
P 507.25 / S 511.75	20	506–512
P 513.25 / S 517.75	21	512–518
P 519.25 / S 523.75	22	518–524
P 525.25 / S 529.75	23	524–530
P 531.25 / S 535.75	24	530–536
P 537.25 / S 541.75	25	536–542
P 543.25 / S 547.75	26	542–548
P 549.25 / S 553.75	27	548–554
P 555.25 / S 559.75	28	554–560
P 561.25 / S 565.75	29	560–566
P 567.25 / S 571.75	30	566–572
P 573.25 / S 577.75	31	572–578
P 579.25 / S 583.75	32	578–584
P 585.25 / S 589.75	33	584–590
P 591.25 / S 595.75	34	590–596
P 597.25 / S 601.75	35	596–602
P 603.25 / S 607.75	36	602–608
P 609.25 / S 613.75	37	608–614
P 615.25 / S 619.75	38	614–620
P 621.25 / S 625.75	39	620–626
P 627.25 / S 631.75	40	626–632
P 633.25 / S 637.75	41	632–638
P 639.25 / S 643.75	42	638–644
P 645.25 / S 649.75	43	644–650
P 651.25 / S 655.75	44	650–656
P 657.25 / S 661.75	45	656–662
P 663.25 / S 667.75	46	662–668
P 669.25 / S 673.75	47	668–674
P 675.25 / S 679.75	48	674–680
P 681.25 / S 685.75	49	680–686
P 687.25 / S 691.75	50	686–692
P 693.25 / S 697.75	51	692–698
P 699.25 / S 703.75	52	698–704
P 705.25 / S 709.75	53	704–710
P 711.25 / S 715.75	54	710–716
P 717.25 / S 721.75	55	716–722
P 723.25 / S 727.75	56	722–728
P 729.25 / S 733.75	57	728–734
P 735.25 / S 739.75	58	734–740
P 741.25 / S 745.75	59	740–746
P 747.25 / S 751.75	60	746–752
P 753.25 / S 757.75	61	752–758
P 759.25 / S 763.75	62	758–764
P 765.25 / S 769.75	63	764–770
P 771.25 / S 775.75	64	770–776
P 777.25 / S 781.75	65	776–782
P 783.25 / S 787.75	66	782–788
P 789.25 / S 793.75	67	788–794
P 795.25 / S 799.75	68	794–800
P 801.25 / S 805.75	69	800–806
P 807.25 / S 811.75	70	806–812
P 813.25 / S 817.75	71	812–818
P 819.25 / S 823.75	72	818–824
P 825.25 / S 829.75	73	824–830
P 831.25 / S 835.75	74	830–836
P 837.25 / S 841.75	75	836–842
P 843.25 / S 847.75	76	842–848
P 849.25 / S 853.75	77	848–854
P 855.25 / S 859.75	78	854–860
P 861.25 / S 865.75	79	860–866
P 867.25 / S 871.75	80	866–872
P 873.25 / S 877.75	81	872–878
P 879.25 / S 883.75	82	878–884
P 885.25 / S 889.75	83	884–890

Why include these in a book devoted to amateur radio? Because by knowing what they are, hams can often prevent their transmitters from causing interference on their own and their neighbors' picture screens.

Until the advent of single sideband equipment, Channel 2 particularly was vulnerable to signals from the ham six-meter band, which covers 50 to 54 megahertz. Note from the first listing in the chart that 54 MHz is also the bottom limit of No. 2. While none of the other bands and channels are that close to each other, it is sometimes possible for secondary signals called harmonics from a slightly mistuned transmitter to cause trouble, especially if high power is used.

In the chart the letter P represents the picture (video) frequency and S the sound (audio) frequency. The figures are in megahertz. Observe that in all cases the sound signal is 4.5 MHz higher than the picture signal. These signals are entirely different in that the picture is amplitude modulated and the sound is frequency modulated. There is on the market a novel little shortwave receiver that takes advantage of this separation to include TV sound on channels 2 through 13 in addition to the regular AM and FM broadcasting stations.

Appendix D
Magazines and Books

Magazines

The amateur radio hobby is well served by seven monthly magazines. The oldest and best known is *QST*, since 1914 the official journal of the hams' national organization, the American Radio Relay League (ARRL). Headquarters and publication office are at 225 Main Street, Newington, Connecticut 06111.

Other magazines worth reading are:

73. 73, Inc., Peterborough, New Hampshire 03458.

Ham Radio and *Ham Radio Horizons*, both from Communications Technology, Inc., Greenville, New Hampshire 03048.

CQ and *Modern Electronics*, both from Cowan Publishing Co., 14 Vanderventer Avenue, Port Washington, New York 10050.

Worldradio, Worldradio Associates, 2120 28th Street, Sacramento, California 95818.

Postcards to these magazines will get you current information on subscription rates.

Books

The ARRL publishes a fine assortment of study pamphlets for people preparing for license tests. These are updated frequently as the FCC rules and regulations change. The League also puts out annual editions of its *Radio Amateur's Handbook* (no relation to this book of the same title!), a monumental tome of 700 pages. The League publications are carried by practically all dealers who sell ham equipment.

RCA Corporation, 30 Rockefeller Plaza, New York, N.Y. 10022, is the source of several highly regarded books of interest to anyone dabbling in electronics. They are:

Receiving Tube Manual, 752 pages.
Transistor, Thyristor & Diode Manual, 656 pages.
Linear Integrated Circuit Fundamentals, 240 pages.
COS/MOS Integrated Circuits Manual, 160 pages.
(COS/MOS means "complementary symmetry/metal-oxide semiconductors."
Hobby Circuits Manual, 224 pages.

Editors and Engineers, a division of Howard W. Sams & Co., 4300 West 62nd Street, Indianapolis, Indiana 46206, publishes Radio Handbook, 974 pages, of particular interest to hams who like to make equipment.

Radio Shack, One Tandy Center, Forth Worth, Texas 76102, issues Dictionary of Electronic Terms; From 5 Watts to 1,000 Watts, a programmed study manual; and All About CB Two-Way Radio.

If you've read all of the foregoing books and thirst for further knowledge of electronics, you might want to write to the following publishers and ask for copies of their current lists of titles on the subject. Also look in your local library and see what you can borrow from its shelves.

Macmillan, Inc., 866 Third Avenue, New York, N.Y. 10022
D. Van Nostrand Company, 450 W. 33rd Street, New York, N.Y. 10001
Pitman Publishing Corporation, 6 Davis Drive, Belmont, Calif. 94002
McGraw-Hill, Inc., 1221 Avenue of the Americas, New York, N.Y. 10020
Prentice-Hill, Inc., Englewood Cliffs, N.J. 07632
John Wiley & Sons, Inc., 605 Third Avenue, New York, N.Y. 10016
Holt, Rinehart & Winston, Inc., 383 Madison Avenue, New York, N.Y. 10017

In the ham magazines, read the advertisements closely for offers of free catalogs and send for them. Many contain technical information of general value.

Appendix E
Addresses of FCC Field Offices

Listed below are the addresses and telephone numbers of the FCC field offices. This list is alphabetical by state, and also includes the field offices in Puerto Rico and the District of Columbia (Washington, D.C.).

ALASKA, Anchorage
U.S. Post Office Building
Room G63
4th & F Street, P.O. Box 644
Anchorage, Alaska 99510
Phone: Area Code 907 265-5201

CALIFORNIA, Long Beach
3711 Long Beach Blvd.
Suite 501
Long Beach, California 90807
Phone: Area Code 213 426-7955

CALIFORNIA, San Diego
Fox Theatre Building
1245 Seventh Avenue
San Diego, California 92101
Phone: Area Code 714 293-5460

CALIFORNIA, San Francisco
323A Customhouse
555 Battery Street
San Francisco, California 94111
Phone: Area Code 415 556-7700

COLORADO, Denver
Suite 2925, The Executive Tower
1405 Curtis Street
Denver, Colorado 80202
Phone: Area Code 303 837-4053

DISTRICT OF COLUMBIA (WASHINGTON, D.C.)
1919 M Street N. W. Room 411
Washington, D.C. 20554
Phone: Area Code 202 632-8834

FLORIDA, Miami
51 S. W. First Avenue, Room 919
Miami, Florida 33130
Phone: Area Code 305 350-5541

FLORIDA, Tampa
809 Barnett Office Building
1000 Ashley Drive
Tampa, Florida 33602
Phone: Area Code 813 228-2605

GEORGIA, Atlanta
440 Massell Building
1365 Peachtree St. N.E.
Atlanta, Georgia 30309
Phone: Area Code 404 881-7381

GEORGIA, Savannah
238 Federal Office Bldg. and Courthouse
125 Bull Street, P.O. Box 8004
Savannah, Georgia 31402
Phone: Area Code 912 232-4321 ext. 320

HAWAII, Honolulu
502 Federal Building
P.O. Box 1021
355 Merchant Street
Honolulu, Hawaii 96808
Phone: Area Code 808 546-5640

ILLINOIS, Chicago
230 South Dearborn Street, Room 3935
Chicago, Illinois 60604
Phone: Area Code 312 353-0195

LOUISIANA, New Orleans
829 F. Edward Hebert Federal Bldg.
600 South Street
New Orleans, Louisiana 70130
Phone: Area Code 504 589-2094

MARYLAND, Baltimore
819 Federal Building
31 Hopkins Plaza
Baltimore, Maryland 21201
Phone: Area Code 301 962-2727

MASSACHUSETTS, Boston
1600 Customhouse
165 State Street
Boston, Massachusetts 02109
Phone: Area Code 617 223-6608

MICHIGAN, Detroit
1054 Federal Building
231 W. Lafayette Street
Detroit, Michigan 48226
Phone: Area Code 313 226-6077

MINNESOTA, St. Paul
691 Federal Building & U.S. Courthouse
316 N. Robert Street
St. Paul, Minnesota 55101
Phone: Area Code 612 725-7819

MISSOURI, Kansas City
1703 Federal Building
601 East 12th Street
Kansas City, Missouri 64106
Phone: Area Code 816 374-5526

NEW YORK, Buffalo
1307 Federal Building
111 W. Huron Street
Buffalo, New York 14202
Phone: Area Code 716 842-3216

NEW YORK, New York
201 Varick Street
New York, New York 10014
Phone: Area Code 212 620-3435

OREGON, Portland
1782 Federal Office Building
1220 S. W. 3rd Ave.
Portland, Oregon 97204
Phone: Area Code 503 221-3097

PENNSYLVANIA, Philadelphia
11425 James A. Byrne Federal Courthouse
601 Market Street
Philadelphia, Pennsylvania 19106
Phone: Area Code 215 597-4410

PUERTO RICO, San Juan
747 Federal Building
Hato Rey, Puerto Rico 00918
Phone: Area Code 809 753-4008-4567

TEXAS, Beaumont
323 Federal Building
300 Willow Street
Beaumont, Texas 77701
Phone: Area Code 713 838-0271

TEXAS, Dallas
Earle Cabell Federal Bldg. & U.S. Courthouse
13E7, 1100 Commerce Street
Dallas, Texas 75242
Phone: Area Code 214 749-3243

TEXAS, Houston
5636 New Federal Office Bldg.
515 Rusk Avenue
Houston, Texas 77002
Phone: Area Code 713 226-4306

VIRGINIA, Norfolk
Military Circle
870 North Military Highway
Norfolk, Virginia 23502
Phone: Area Code 804 461-4000

WASHINGTON, Seattle
3256 Federal Building
915 Second Ave.
Seattle, Washington 98174
Phone: Area Code 206 442-7610

Glossary

A – (*A* **negative**). Symbol usually used to designate the negative terminal of a direct-current filament source.

A + (*A* **positive**). Symbol usually used to designate the positive terminal of a direct-current filament source.

A **battery.** The battery used to supply power to the filaments of vacuum tubes.

absorption wavemeter. A device for measuring wavelength or frequency. Usually consists of a tunable resonant circuit and an indicator to show when maximum energy is being absorbed from circuit being tested.

accelerating electrode. One or more internal elements of an electron tube used to increase the velocity of an electron stream.

AC-DC. Term applied to electronic equipment that can be operated from either alternating or direct current.

acoustic. Pertaining to the generation, transmission, and effects of sound.

ADF, adf, A.D.F., a.d.f. Automatic direction finder.

adjustable resistor. A resistor whose value can be changed mechanically. Also adjustable voltage divider.

admittance. The measure of the ease with which current flows in a circuit: the reciprocal of resistance. Admittance is measured in mhos and designated by *Y*.

aerial. A system of electrical conductors used for reception or transmission of radio waves. The word antenna is generally used in the same sense.

AFC. Automatic frequency control.

air core. Descriptive term for coils or transformers with air cores, used chiefly in radio-frequency circuits.

align. To adjust circuits so that they function properly.

aligning tool. Small screwdriver or special tool, generally of non-inductive material, for aligning circuits.

Allen screw. Screw having recessed hexagonal keyway in the head.

alligator clip. A long-nosed metal clip with meshing jaws, used to make temporary connections.

all-wave antenna. An antenna designed to receive or radiate over a wide range of radio frequencies.

all-wave receiver. A set capable of broadcast and shortwave reception. Usually tunes from about 500 kc to 30 MHz. The term general-coverage is used in the same sense.

alnico. Permanent magnet alloy of iron with aluminum, cobalt, and nickel for loudspeakers, motors, meters, etc.

alternating current. A term used to distinguish current of changing polarity from direct or constant polarity current.

amateur. The term for licensed operators whose aims are public service and technical experimentation. More familiarly, they are "hams."

amateur bands. Radio-frequency bands assigned to amateurs by international agreement.

amateur station call sign. The identification assigned to an amateur station by the Federal Communications Commission.

American Morse Code. A system of dots and dashes for telegraphy (never used in radio but commonly over telegraph lines). In radio, International Morse Code is used.

American Wire Gauge. Standard system for measuring wire diameters. Abbreviated AWG.

ammeter. An instrument for measuring current flow in amperes.

ampere. The unit of current flowing through 1 ohm resistance at 1 volt potential in 1 second. Abbreviated amp.

ampere-hour. 1 ampere of current flowing for 1 hour.

amplification. The process of increasing the strength of audio- and radio-frequency currents.

amplification factor. Rating applied to vacuum tubes to indicate the maximum increase in signal strength theoretically available with a given tube.

amplitude. Term used to describe the magnitude of a wave; the largest or peak value measured from zero.

amplitude modulation. The modulating of a carrier-frequency current by varying its amplitude above and below normal value at an audio rate, in accordance with the voice or music being transmitted. Common abbreviation is AM.

analyzer. A test instrument for checking electronic equipment, parts, or circuits.

angle of lag or lead. Phase angle by which voltages or currents precede or follow one another. These relations are often indicated by plotting the sinusoidal curves along an axis of electric degrees. They may also be pictured by vectors.

angle of radiation. The angle between the center of a radiated radio beam and the earth's surface.

anode. The element of a vacuum tube or corresponding solid-state device toward which the internal electron stream flows when it is made positive. In a tube this element is the plate. See also the word cathode.

antenna. See aerial.

antenna coil. The RF coil (or transformer) in a radio receiver or transmitter to which the antenna is connected.

antenna coupler. A device used to connect a receiver or transmitter to an antenna or antenna transmission line.

antenna current. The current flowing in the antenna and associated circuits.

antenna transmission line. A system of conductors connecting an antenna to a receiver or transmitter, often through an antenna coupler.

A-power supply. Power source for the filaments of electron tubes.

arc. A luminous discharge resulting from the passage of electric current across a path of ionized air, vapor, or gas.

armature. The moving portion of an electromagnetic circuit, such as the rotating section of a generator or motor.

array. A combination of antenna elements usually arranged so that each element reinforces the performance of the other. An array is often used when very high directivity and gain are required.

ARRL. American Radio Relay League, an organization of radio amateurs.

atmospheric interference. Crackling and hissing noises reproduced by a receiver as a result of electric disturbances in the atmosphere. Also called static.

atom. Smallest unit of any chemical element. Atoms consist of systems of fundamental particles: protons, neutrons, and electrons, arranged with a characteristic structure for each element.

attenuation. Reduction in the strength of currents.

audible. Capable of being heard by the human ear.

audio. Pertaining to voltages or currents in the audible frequency range.

audio amplifier. A device to strengthen audio-frequency signals.

audio frequency. A frequency in the range of audible sound waves: 15 to 20,000 hertz.

audio-frequency oscillator. A device which generates audio-frequency signals.

audio transformer. An iron-core transformer used in audio-frequency circuits.

autodyne reception. Radio reception in which the incoming signal beats with an oscillating detector to produce an audible beat frequency; employed in regenerative receivers for the reception of CW (continuous wave) code signals.

automatic frequency control. A circuit which keeps a receiver or transmitter accurately tuned to a predetermined frequency. Abbreviated AFC.

automatic volume control. A circuit which automatically maintains a constant output volume in spite of varying input signal. Used in practically all modern receivers where it minimizes fading and prevents blasting when tuning suddenly from a weak station to a strong one. Abbreviated AVC.

autotransformer. Any single-coil transformer in which the primary and secondary connections are made to the single coil.

B. Letter normally used to designate the high-voltage plate power supply for vacuum tubes.

B − (*B* **negative**). Symbol used to designate the negative terminal of the plate supply.

B + (*B* **positive**). Symbol used to designate the positive terminal of the plate supply.

back-electromagnetic force. Abbreviated back-emf. Also called counterelectromotive force. A voltage created in an inductive circuit by an alternating current flow. The polarity of the back-emf is opposite to that of the applied voltage.

balance to ground. A state in certain circuits (e.g., cathode-ray tubes) where the voltages (such as on deflection plates) are equal above and below ground potential.

ballast resistor. A special type of resistor used to compensate for fluctuations in AC power line voltage. The resistance of a ballast resistor increases as the current through it increases, thus maintaining the current essentially constant in spite of line voltage fluctuations.

ballast tube. A ballast resistor mounted inside an evacuated envelope.

banana jack. A receptacle that fits a banana plug.

banana plug. A banana-shaped plug. Elongated springs provide compression contact.

band. In radio, frequencies which are within two definite limits. For example, the standard broadcast band extends from 550 to 1600 kc.

band-pass coupling. A type of coupling between stages that provides relatively linear energy transfer over a wide band of frequencies.

band-pass filter. A filter that passes a specified frequency band while all frequencies above and below this band are attenuated.

bandspread. Any method, mechanical or electronic, of effectively widening the tuning scale of a receiver between radio stations.

band switch. Used to change one or more circuits of a multiband radio receiver or transmitter; also called band selector.

bandwidth. A section of the frequency spectrum required to transmit the desired intelligence. For example, the bandwidth of the average AM broadcast channel is 10 kc.

base insulator. A large insulator used at the base of some radio towers.

bathtub capacitor. A capacitor enclosed in a metal can with rounded corners like a bathtub.

battery. Two or more dry cells or storage cells connected together to act as a DC voltage source. A single cell is loosely called a battery.

B **battery.** Used to supply DC to the plates and screens of vacuum tubes.

BC. Broadcast band.

BCI. Broadcast interference. Term used to denote interference by amateur transmitters with reception of broadcast signals on standard broadcast receivers.

BCL. Broadcast-band listener (as distinguished from an amateur operator).

beam. (1) A colloquialism for an antenna array. (2) A stream of electrons flowing from the cathode to the plate in beam power tubes. (3) A stream of electrons flowing from the cathode to the screen of a television or cathode ray tube.

beam antenna. Antenna array that receives or transmits radio-frequency energy more sharply in one direction than others.

beat frequency. The frequency obtained when signals of two different frequencies are combined; equal in numerical value to the sum and difference of the original frequencies.

beat-frequency oscillator. A device from which an audible signal is obtained by combining and rectifying two higher inaudible frequencies. Abbreviated BFO.

beat reception. Radio reception by combining a received external signal with an internal one generated in the receiver; the difference frequency is then amplified and detected. Also referred to as heterodyne reception.

bias. Fixed DC voltage applied between the cathode and grid of vacuum tubes and the emitter and base of transistors to control the amplifying action of the device.

bias modulation. A means of amplitude modulation in which the modulating voltage is superimposed on the bias voltage of an RF stage. Control grid, suppressor grid, and cathode modulation are types of bias modulation.

bias resistor. The cathode resistor through which tube current flows, to develop a DC voltage used as a C bias.

binding post. A fixed terminal to which wires may be attached.

blanking. The cutting off of the beam in a cathode-ray tube during a desired interval, such as when the spot is rapidly returning to begin a new sweep in an oscilloscope.

bleeder current. A current drawn continuously from a power supply to improve its voltage regulation.

bleeder resistor. A resistor used to draw a fixed bleeder current from a power supply. Acts as a safety device by discharging filter condensers after the power supply is turned off.

block diagram. Simplified outline of an electronic system with circuits or parts shown as functional boxes.

blocked-grid keying. A means of keying a radio-telegraph transmitter by means of a blocking bias on the control grid of one or more tubes.

blocking. The application of high negative grid bias to a vacuum tube, reducing tube current to zero.

blocking capacitor. Any capacitor used in a circuit to block the flow of direct current while allowing AC signals to pass through.

body capacitance. The capacitance existing between the human body and a piece of radio equipment.

breadboard. Idiom for an experimental circuit setup on a board.

break-down voltage. The voltage at which the insulation between two conducting elements will conduct an appreciable amount of current.

bridge circuit. Consists of four arms in a diamond-shaped arrangement. When they are simple diodes, the circuit provides full-wave rectification of AC in power-supply equipment. When the arms consist of two fixed resistors and one variable, all of known value, and another of unknown value, they constitute a Wheatstone bridge, by means of which the value of the fourth resistor can be measured quickly and with great accuracy.

broadband. Ability of a circuit or antenna to be effective over a relatively wide frequency range.

broadband amplifier. An amplifier that maintains flat response over a relatively wide range of frequencies.

broadband antenna. A transmitting or receiving antenna that is uniformly efficient over a relatively wide frequency band.

broadband RF stage. An amplifier stage that provides approximately uniform amplification over a wide band of frequencies.

broadcast band. The band of frequencies between 550 and 1600 kc in which are assigned all standard AM broadcast stations operating in the United States.

broadcasting. A general term applying to radio transmission of material for general public listening.

B supply. The plate voltage source for vacuum tubes.

buffer. Any part or circuit used to reduce undesirable interaction between two or more circuits.

bug. A semiautomatic code-transmitting key in which movement of a lever to one side produces dots, and movement to the other side produces dashes.

bulb. Glass or metal shell of a vacuum tube; sometimes called envelope.

bus. Term used to specify an uninsulated conductor (a bar or wire).

butterfly capacitor. A variable capacitor whose plates roughly approximate the shape of a butterfly.

bypass capacitor. Used to provide a low-impedance path for radio or audio signals around a circuit or to ground.

C − (C negative). Symbol used to designate the negative terminal of the grid bias voltage source.

C + (C positive). Symbol used to designate the positive terminal of the grid bias voltage source.

cable. One or more insulated or noninsulated conductors. Grouped insulated conductors. Grouped insulated wires are called a multiconductor cable.

calibration. A method of comparing an instrument, device, or dial with a standard to determine its accuracy.

capacitance. The electric charge that can be received by a system of insulated conductors from a potential source. The unit of capacitance is the farad.

capacitance bridge. A variant of the Wheatstone bridge used to make exact comparisons of capacitances.

capacitive coupling. Coupling in which a capacitor provides a path for signal energy between two circuits or stages.

capacitive reactance. The reactance which a capacitor offers to AC or pulsating DC. It is measured in ohms, and decreases as frequency and capacitance are increased.

capacitor. A device having the property of capacitance. Sometimes called a condenser.

capacitor-input filter. A type of power-supply filter in which a capacitor precedes an inductor or a resistor across the output of the rectifier.

carbon microphone. A type of microphone in which the pressure of sound waves against a diaphragm is transmitted mechanically to a number of carbon granules, thereby causing their resistance to vary in accordance with the sound impressed on the diaphragm.

carbon resistor. A resistor made of carbon particles and a ceramic binder molded into a cylindrical shape, with axial leads.

carrier. A current, voltage, or radio wave at the assigned frequency of a radio station.

carrier suppression. Radio transmission in which the energy of the carrier wave is greatly reduced.

cascade. Actually, in series; as in amplifier stages where the output of one stage is connected to the input of the next.

cathode. The negative or the electron-emitting electrode of a vacuum tube, indirectly heated by a filament located inside the cathode, or directly heated by current flowing through the cathode itself, in which case the filament is the cathode itself. Gas tubes employ cold cathodes.

cathode follower. A vacuum-tube stage wherein the output is taken between cathode and ground, providing high-impedance input with low-impedance output.

cathode keying. Method of keying a transmitter by opening the plate return lead to the cathode or filament center tap.

cathode modulation. Amplitude modulation by varying the cathode bias of an RF amplifier in accordance with the modulating intelligence.

cathode-ray oscilloscope. A test instrument using a cathode-ray tube, providing a visible graphic presentation of electrical impulses.

cathode-ray tube. A type of funnel-shaped tube in which a beam of electrons generated at the apex of the tube impinges on a fluorescent screen on the inner face of the tube, thereby causing a spot of light on the face. Voltages applied to vertical and horizontal pairs of deflection plates control the position of the beam and hence the spot on the face of the tube. Used in oscilloscopes and television receivers.

cathode-ray tube screen. The fluorescent material (phosphor) that covers the inside surface of the face end of a cathode-ray tube.

cathode-ray tuning indicator. A very small cathode-ray tube used in receivers to indicate when a station is properly tuned.

C battery. The battery used for supplying a negative bias potential to the control grid of a vacuum tube.

center-fed antenna. A type of transmitting or receiving antenna having a transmission line connected to its center.

center frequency. The assigned frequency of an FM station; frequency shifts take place in step with the audio signal.

ceramic. A material composed of aluminum and magnesium oxides, which after molding and firing is used as insulation. It withstands high temperatures and is less fragile than glass.

ceramic capacitor. A capacitor with a ceramic dielectric.

channel. (1) A band of frequencies including the assigned carrier frequency, within which transmission is confined in order to prevent interference with stations on adjacent channels. (2) An electrical path over which signals travel; thus, an amplifier may have several input channels, such as microphone, tuner, or phonograph.

chassis. The metal framework on which electronic components are mounted.

choke coil. An inductor which resists the flow of alternating current while allowing direct current to pass. Radio-frequency choke coils have air or pulverized-iron cores, while AF and filter chokes have laminated sheet-iron cores.

choke input filter. Network of capacitors and inductors, the first member being a choke.

circuit breaker. A device for opening a circuit should the current exceed a predetermined value.

click filter. A network which reduces or eliminates the key clicks in a radiotelegraph transmitter.

clipping. Distortion in amplifiers produced by flattening the positive and/or negative peaks of the signal due to tube saturation during positive grid swing, or due to driving the grid below cutoff. Also, distortion in the AF component of a modulated wave when modulation amplitude exceeds 100 percent.

coaxial cable. Also coax cable or coax. A two-conductor cable in which one conductor is a flexible or nonflexible metal tube and the other is a wire axially supported inside the tube by insulators.

code. A system of dot and dash signals used in the transmission of messages by radio or wire telegraphy. The International Morse Code (also called the Continental Code) is used universally for radiotelegraphy.

cold cathode tubes. Tubes in which cathodes are not heated. These include vacuum tubes such as photoelectric cells and rectifiers and gas glow tubes such as voltage regulators.

color code. Any system of colors used to specify the electric value of a radio part, terminals, or leads.

Colpitts oscillator. A popular type of oscillator employing capacitive feedback to sustain oscillation.

conductor. A material which offers little opposition to the continuous flow of electric current.

cone (*speaker*). The conical-shaped paper or fiber diaphragm of a loudspeaker.

connector. A device for interconnecting one or more cables or electronic circuits. There are two main classifications of connectors: male and female. A female connector has contacts set in recessed openings. These openings accommodate a male connector to make electrical connection. A line cord plug is a simple type of male connector; a wall electric outlet is a simple type of female connector. Male connectors are often called plugs; female connectors are often referred to as jacks, sockets, or receptacles.

contact microphone. A crystal device which converts mechanical vibrations into electrical currents. Used industrially and to amplify stringed musical instruments.

contact resistance. The resistance, measured in ohms, between the contacts on a switch or relay.

continuous wave. Abbreviation, CW. An unmodulated, constant-amplitude radio-frequency wave.

control grid. That electrode in a vacuum tube which has the most effective control over the plate current passed by the tube. The control grid is usually the electrode nearest the cathode.

converter. The circuit in a super-heterodyne radio receiver which changes incoming signals to the intermediate frequency; the converter section includes the oscillator and the first detector. Also, a device changing electric energy from one form to another, such as AC to DC, etc.

copper-oxide rectifier. A rectifier made up of disks of copper coated on one side with cuprous oxide.

coulomb. The charge, or quantity of electricity, delivered by a current of 1 ampere flowing for 1 second.

counter emf. A flow of current in an inductor building up a voltage which tends to flow in the opposite direction to the impressed voltage, opposing the original current flow. Also known as counterelectromotive force.

coupling. The means by which signals are transferred from one radio circuit to another. Coupling can be direct through a conductor, electrostatic through a capacitor, or inductive through a transformer.

C power supply. Source of bias voltage for a vacuum tube.

critical coupling. The closest coupling of two circuits tuned to the same frequency at which there will be only one resonant peak.

CRO. Cathode-ray oscilloscope.

cross-modulation. Interference which modulates a signal undesirably, usually from some unwanted source.

crystal. A piece of quartz or similar piezoelectric material which has been ground to proper size to produce natural vibrations at a desired radio frequency. Quartz crystals are used in radio transmitters to generate the assigned carrier frequency, and also as very high Q filters.

crystal-controlled converter. A type of radio-frequency converter employing a piezoelectric crystal to establish the frequency of its oscillator.

crystal-controlled oscillator. A type of radio-frequency oscillator employing a piezoelectric crystal to establish its operating frequency. Oscillators of this type provide high stability.

crystal-controlled transmitter. A transmitter employing a crystal-controlled oscillator to establish its carrier frequency.

crystal microphone. A microphone whose output voltage is created by the deformation of a piezoelectric crystal when subjected to sound wave compression.

current. The movement of electrons through a conductor, measured in amperes.

current feed. The feeding current to an antenna at the point of maximum current amplitude.

current-limiting resistor.   A resistor used in a circuit as a protective device for tubes and filter elements against overload from voltage surges.

cutoff frequency.   That frequency in a filter or other system at which rapid attenuation takes place.

cutoff voltage.   Negative grid bias beyond which plate current ceases to flow.

CW.   Continuous wave. In its amateur application, this means radio-telegraphy by means of the International Morse Code.

DC generator.   A rotating device that converts mechanical energy into unidirectional electric energy.

DC plate resistance.   In ohms, the DC plate voltage divided by the DC plate current of a vacuum tube.

DC resistance.   In ohms, the opposition to current flow offered by a circuit or component to DC current flow.

decibel (db).   A term expressing a ratio between two amplitudes or energies. The db unit between two amplitudes is computed as twenty times the log of the ratio; between two energies as ten times the log of the ratio. Practically, a decibel is approximately the smallest change in sound intensity that the human ear can detect.

decoupling circuit.   A network of one or more resistors and capacitors that separate and bypass unwanted signals.

deflecting electrodes.   Pair of cathode-ray tube electrodes to which the electron beam moves in a horizontal or vertical direction, depending upon the applied potentials.

degenerative feedback.   See negative feedback.

delta connection.   Connection that forms a triangle like the Greek letter delta Δ.

demodulation.   The reverse of modulation. The process of extracting the modulating intelligence, commonly called detection.

detector.   Commonly, the stage or circuit in a radio set that demodulates the RF signal. Generally, a device or circuit that changes the form of a recieved signal into a more usuable form for a specific purpose.

diaphragm.   A thin, flexible sheet which vibrates when struck by or when producing sound waves, as in a microphone, earphone, or loudspeaker.

dielectric.   The insulating material between the plates of a capacitor or adjacent wires in a cable, or between any two conducting elements.

dielectric constant. The relative permittivity of the dielectric material as compared to vacuum. That of air is 1; transformer oil has a dielectric constant of about 2.

dielectric loss. Energy loss in the dielectric of a capacitor. It shows up as heat.

dielectric strength. The maximum voltage that a dielectric can withstand without breakdown. Expressed in volts per millimeter. The dielectric strength of air is 4,000, of mica 50,000.

diffraction. The bending of waves (light, sound, or radio) around the edges of obstacles.

diode. A component having two electrodes, one the cathode and the other the anode.

diode detector. A diode used in a demodulation circuit. Detection may be half- or full-wave rectification.

dipole antenna. A conductor one-half wavelength long at a given frequency; used to radiate or pick up radio waves.

direct coupling. The use of a conductor to connect two amplifier stages together and provide a direct path for the signal currents.

direct current. An unchanging electric current which flows in only one direction.

directional antenna. Any antenna which picks up or radiates signals better in one direction than another.

direction finder. See radio direction finder.

director. An auxiliary antenna element located in front of the radiating or main receiving antenna element so that radiation or reception will be strengthened in the forward direction.

direct resistance-coupled amplifier. An amplifier in which the plate of one stage is connected either directly or through a resistor to the control grid of the next stage.

direct wave. A radio wave that travels directly from transmitting antenna to receiving antenna without being reflected or refracted.

dissipation. Unused or lost energy.

distortion. Unfaithful reproduction of signals due to changes occurring in the wave form of the original signal.

distress signal. By code: SOS; by radiotelephone: "Mayday," a shortened phonetic form of the French expression "prière de m'aider."

distributed capacitance. Capacitance distributed between wires, between parts, between conducting elements themselves, or between the elements and ground, as distinguished from capacitance concentrated or lumped in a capacitor.

distributed inductance. The inductance that exists along the length of a conductor, as distinguished from inductance concentrated in a coil.

double-conversion superheterodyne. A superheterodyne receiver using two first detectors and two IF frequencies.

double diode. Two diodes in the same envelope. Also called duodiode.

double-pole switch. A switch which simultaneously opens or closes two separate circuits or both sides of the same circuit.

doubler. In a transmitter, a circuit in which the output is tuned to twice the frequency of the input circuit. Also power supplies that double the voltage.

doublet antenna. An antenna composed of two elements usually strung in a straight line and connected at the middle to a single insulator, each element being some definite fraction or whole of the desired wavelength.

double-throw switch. A switch which connects one set of terminals to either of two other sets of terminals.

double triode. Two triodes in the same tube envelope. Also called duotriode.

drain. A term used to indicate the current taken from a voltage source.

driver stage. The amplifier stage preceding the high-power audio-frequency or radio-frequency output stage.

driver tube. The tube used in a driver stage.

drop. Voltage drop, due to current flow through an electronic component.

dropping resistor. A resistor used to decrease the voltage in a circuit.

dry cell. A type of primary cell in which the electrolyte is in the form of a paste rather than a liquid.

dry electrolytic capacitor. An electrolytic capacitor in which the electrolyte is a paste.

dual capacitor. Two capacitors in a single housing.

dummy antenna. A resistor or other device which duplicates the characteristics of a transmitting antenna without radiating signals. For testing and adjusting transmitters.

DX. Colloquialism for word "distance," referring to distant reception of radio signals.

dynamic loudspeaker. A loudspeaker in which the coil carrying the audio-frequency current is attached to a moving diaphragm or cone, and moves in and out in a constant magnetic field.

dynamic microphone. A microphone in which the flexible diaphragm is attached directly to a coil positioned in the fixed magnetic field of a permanent magnet.

E_g. Symbol for grid bias voltage.

E_p. Symbol for DC plate voltage.

E_{sg}. Symbol for DC screen-grid voltage.

earphone. See headphone.

e.c. Enamel-covered wire.

echo. Simultaneous reception of a radio signal and the part of it which was transmitted approximately $1/7$ of a second earlier and has circled the earth. The characteristic "hollow" sound, or echo, is produced because of the very small time interval involved.

ECO. Electron-coupled oscillator.

eddy currents. Circulating currents induced in conducting materials by varying magnetic fields. They are usually undesirable because they represent loss of energy and cause heating. Eddy currents are kept at a minimum by employing laminated, powdered, or sintered construction for the iron cores of transformers, AF choke coils, and other magnetic devices. Eddy currents are useful as the source of heat in induction furnaces.

E **layer.** An ionized layer in the E region of the ionosphere.

Electralloy. An alloy frequently used to make radio chassis. It is characterized by having non-magnetic properties.

electric angle. A method of indicating a particular instant of segment in an alternating-current cycle. One cycle is considered equal to 360 degrees, hence $1/2$ cycle is 180 degrees and $1/4$ cycle is 90 degrees.

electric degree. One-360th of a cycle of an alternating current or voltage.

electric eye. A cell or vacuum tube that produces current when light shines on it.

electrode. One of the elements inside a vacuum tube, such as the cathode, plate, or grid.

electrolysis. The production of chemical changes by passing current through an electrolyte.

electrolyte. The liquid, chemical paste, or other conducting medium used between the electrodes of a dry cell, storage cell, electrolytic rectifier, or electrolytic capacitor.

electrolytic capacitor. A fixed capacitor in which the dielectric is a paste electrolyte.

electromagnet. A coil of wire wound on an iron core which produces a strong magnetic field when current is applied.

electromagnetic energy. Energy in a radio wave.

electromagnetic spectrum. Range of frequencies of electromagnetic waves.

Type of Wave	Wavelength
Radio	Above 1,000 km to below 1 cm
Infrared (heat rays)	0.03 to 0.000076 cm
Visible light	0.000076 to 0.000040 cm
Ultraviolet	0.000040 to 0.0000013 cm
X rays	10^{-6} to 10^{-9} cm
Gamma rays	10^{-8} to 5×10^{-11} cm
Cosmic rays	10^{-11} to 10^{-12} cm

electromagnetic waves. Radiation taking many different forms, but all having in common the characteristic of a velocity that is just about 3×10^{10} cm/sec.

electromagnetism. Magnetic effects produced by electric currents.

electromotive force. Voltage.

electron. The elementary charge of negative electricity. The charge is -1.6×10^{19} coulombs.

electron coupling. A process of coupling two circuits through a vacuum tube, principally multigrid tubes.

electron gun. Tube electrodes designed for the production of a narrow beam of electrons intended for use in fluorescent screen or microwave tubes.

electrostatic. Pertaining to electricity at rest.

electrostatic charge. An electric charge stored in a capacitor or on the surface of an insulated object.

electrostatic field. The region near an electrically charged object.

electrostatic shield. A grounded metal screen, sheet, or conductor placed to prevent the action of any electric field through the shield.

emission. (1) The process of radiating radio waves into space by a transmitter. (2) The process of ejecting electrons from a heated material.

end effect. (1) The effect of capacitance at the ends of an antenna. (2) The effect of inductance at the end of a coil.

envelope.   The glass or metal housing of a vacuum tube.

***E* region.**   A region in the ionosphere, from about 55 to 85 miles above the surface of the earth, containing ionized layers capable of bending or reflecting radio waves.

excitation.   Application of a signal to the input of a vacuum tube or similar solid-state device. Or, application of voltage to the field coils of a motor, generator, loudspeaker, or other device that requires a magnetic field.

exciter.   The oscillator that generates the carrier frequency of a transmitter.

F.   Symbol for filaments of a tube.

facsimile.   A system of communication in which previously reproduced images, such as photographs or printed matter, are transmitted graphically.

fade.   To change gradually in signal amplitude.

farad.   Unit of capacitance. In the practical system of units, the farad is too large for ordinary use; so, measurements are made in terms of microfarads and picofarads.

Federal Communications Commission (FCC).   A board of commissioners appointed by the President, having the power to regulate all communications systems originating in the United States.

feedback.   Transfer of a portion of energy from one point to a preceding point.

feedthrough capacitor.   A very efficient type of bypass capacitor, designed so that the inner foil is in series with the wire to be bypassed and the outer shell is coaxial to the wire.

feedthrough insulator.   A type of insulator which permits feeding of wire or cable through walls, etc., with minimum current leakage.

fidelity.   The degree of exactness with which a system or portion of a system reproduces an input signal.

field.   The effect produced in surrounding space by an electrically charged object, by electrons in motion, or by a magnet.

field coil.   An insulated winding energized by DC voltage, and mounted so as to magnetize a field pole.

field pattern.   Usually expressed as a polar diagram, indicating the horizontal field strength of an antenna.

field-strength meter.   A measuring instrument used to determine the strength of radiated energy (field strength) from a radio transmitter.

filament. The wire through which current is sent in a vacuum tube to produce the heat required for electron emission.

filament emission. Evolution of electrons from a heated filament in a vacuum tube.

filament winding. A separate secondary winding on the power transformer of AC-operated apparatus used as a filament voltage source.

filter. A resistor, coil, capacitor, or any combination of such parts used to block or attenuate alternating currents at certain frequencies while allowing essentially unimpeded flow of currents at other frequencies.

filter capacitor. A capacitor used in a power-pack filter system to provide a low-reactance path for alternating currents.

filter choke. An iron-core coil used in a power-pack filter system to pass direct current while offering high impedance to pulsating or alternating currents.

first detector. The stage in a superheterodyne receiver in which the beat-frequency signal is combined with the incoming radio-frequency signal to produce the intermediate-frequency signal. Also called a mixer.

flashover. A disruptive discharge over the surface of an insulator, or between two charged surfaces not in mutual contact.

fluorescent. Having the property of giving off light when activated by electronic bombardment.

fluorescent screen. A sheet of suitable material coated with a phosphor that fluoresces visibly when hit by an electron beam.

flux. (1) A material used to promote fusion or joining of metals in soldering. Rosin is widely used as a flux in electronic equipment. (2) A term used to designate collectively the magnetic lines of force in a region.

flux density. The number of electric or magnetic lines of force cutting a unit area at right angles.

flywheel effect. The effect of a resonant circuit. Although the grid controls the input energy in pulses as in the cylinder explosions of a gasoline engine, the resonant circuit maintains continuous operation as does the flywheel of the engine.

free electron. An electron which is not attached to any one atom, but is free to move from atom to atom.

F region. That region of the ionosphere extending from about 90 to 250 miles above the earth's surface.

frequency. The number of complete cycles or vibrations per unit of time, usually per second. Frequency of a wave is equal to the velocity divided by the wavelength.

frequency conversion. The process of converting the frequency of a signal to some other frequency by combining it with another frequency.

frequency discriminator. A circuit that converts a frequency-modulated signal into an audio signal.

frequency doubler. A vacuum-tube stage having a resonant plate circuit tuned to twice the input frequency.

frequency drift. An undesired change in the frequency of an oscillator, transmitter, or receiver.

frequency modulation. A method of modulating a carrier frequency by causing the frequency to vary above and below a center frequency in accordance with the sound to be transmitted. The amount of deviation in frequency above and below the center frequency is proportional to the amplitude of the modulating intelligence. The number of complete deviations per second above and below the center frequency corresponds to the frequency of the modulating intelligence.

frequency multiplier. A frequency changer used to multiply a frequency by an integral value, such as a frequency doubler.

frequency response. A rating or graph which expresses the manner in which a circuit or device handles the different frequencies falling within its operating range.

frequency shift. A deliberate change in the frequency of a radio transmitter or receiver.

frequency-shift transmission. A system of automatic code transmission that shifts the carrier frequency back and forth between two frequencies instead of keying the carrier on and off.

frequency stability. The ability of an oscillator to maintain a predetermined frequency.

frequency standard. An oscillator used for frequency calibration.

front-to-back ratio. The ratio of the effectiveness of a directional antenna or microphone between the front and the rear.

fundamental. The lowest frequency component of a complex vibration, tone, or electric signal.

fuse. A protective device consisting of a short piece of wire which melts and breaks when the current through it exceeds the rated value.

G. Symbol for control grid of a tube.

g$_m$. Designation for the mutual conductance of a vacuum tube.

gain. The ratio of output voltage, current, or power to the input voltage, current, or power in an amplifier stage or system. Usually expressed in db.

gain control. A control that can change the overall gain of an amplifier. A volume control.

galvanometer. A D'Arsonval laboratory instrument for measuring or indicating extremely small electric currents.

gang capacitor. Two or more variable capacitors mechanically mounted so that they can be simultaneously turned by a single shaft.

gang control. A number of similar pieces of apparatus that can be simultaneously adjusted or tuned by a single control or shaft.

gap arrester. A type of antenna lightning arrester employing one or more air gaps connected between the antenna and ground.

gaseous rectifier. A gas-filled rectifier. May have a hot cathode, a mercury pool, or a cold cathode.

germanium. A grayish-white brittle metallic element widely used in transistors.

getter. An alkali or alkaline earth metal introduced into a vacuum tube during manufacture and vaporized after the tube has been evacuated, to absorb any gases which may have been left by the vacuum pump.

glow-discharge tube. A tube which conducts by ionization of a gas such as a neon tube.

glow-discharge voltage regulator. A gas tube that maintains a nearly constant voltage when connected across a mildly varying voltage source. Often called "VR tube."

grid. An electrode mounted between the cathode and the anode of a vacuum tube to control the flow of electrons from cathode to anode.

grid bias. The DC voltage applied to the control grid of a vacuum tube to make it negative with respect to the cathode.

grid-bias cell. A small cell used in the grid circuit of a vacuum tube to provide C bias voltage.

grid circuit. The circuit between the grid and cathode of a vacuum tube. The input circuit of the tube.

grid clip. A spring clip to make a connection to the top (grid) cap terminal on some vacuum tubes.

grid current. The current passing to or from a grid through space inside a vacuum tube.

grid detection. Detection taking place due to the action of the grid circuit of a vacuum tube, as in a grid-leak detector.

grid-dip meter. An oscillator having in its input circuit a sensitive current-indicating meter that dips when energy is drawn from the oscillator by a coupled resonant circuit.

grid driving power. The wattage applied to the grid circuit of a tube.

grid leak. A resistor used in the grid circuit of a vacuum tube to provide a discharge path for grid current.

grid modulation. Modulation produced by introduction of the modulating intelligence into the grid circuit.

grid-plate capacitance. The capacitance between the grid and the plate within a vacuum tube.

grid return. The lead or connection which provides a path for electrons from the grid circuit to the cathode.

grid swing. The total grid signal voltage variation from the positive to negative peaks.

grid voltage. The voltage between grid and cathode.

grommet. An insulating washer, usually made of rubber or a plastic material, used to prevent a wire from touching a chassis or panel.

ground. A connection, intentional or accidental, between a live circuit and the earth or some conducting body serving as the earth.

ground absorption. Transmitted radio power dissipated in the ground.

grounded. Connected to earth or to some conducting body that serves as the earth.

grounded-grid amplifier. A circuit in which the input is applied to the cathode rather than to the grid of a triode tube.

ground-return circuit. A circuit that is completed by utilizing the earth as a conductive path.

ground wave. A radio wave that is propagated near or at the surface of the earth.

ground wire. A conductor leading to an electrical connection with the ground.

half-wave antenna. An antenna whose length is approximately equal to one-half the wavelength to be transmitted or received.

half-wave line. A transmission line having an electric length equal to one-half the wavelength of the signal to be transmitted or received.

half-wave rectification. Rectification of only one-half of each alternating-current cycle into pulsating direct current.

half-wave rectifier. A radio tube or other device which converts alternating current into pulsating direct current by allowing current to pass during one-half of each alternating current cycle.

ham. A term applied to licensed amateur radio operators.

harmonic. A sinusoidal wave that is an integral multiple of the fundamental frequency, which is called the first harmonic.

harmonic content. The degree or numbers of harmonics in a complex frequency output.

harmonic generator. A vacuum tube or other generator which produces an alternating current having many harmonics.

harmonic suppression. The prevention of harmonic generation in an oscillator or in circuits that follow it.

harness. Wires and cables so arranged and tied together that they may be connected or disconnected as a unit.

Hartley oscillator. A vacuum-tube oscillator circuit identified by a tuned circuit employing a tapped winding connected between the grid and plate.

headphone. A telephone receiver, used either singly or in pairs.

headset. A pair of headphones attached to a headband to hold the phones against the ears.

heater. An electric heating element for supplying heat to an indirectly heated cathode in an electron tube.

heater current. The current flowing through a heater serving an indirectly heated cathode.

heater voltage. The voltage between the terminals of a filament used for supplying heat to an indirectly heated cathode.

Heaviside layer. A layer of ionized gas in the region between 50 and 400 miles above the surface of the earth which reflects radio waves back to earth under certain conditions.

henry. The practical unit of self- or mutual inductance. The inductance in which a current changing its rate of flow 1 ampere per second induces an electromotive force of 1 volt.

hertz. One complete alternating-current reversal, consisting of rise to a maximum in one direction, a return to zero, a rise to a maximum in the other direction, and another return to zero. The number of hertz during one second is the frequency of an alternating current.

heterodyne. Pertaining to the production of a frequency (beat) by combining two frequencies.

heterodyne frequency. The beat frequency, which is the sum or difference frequency of two signals.

heterodyne reception. The process of receiving radio waves by combining a received radio-frequency voltage with a locally generated alternating voltage to produce a beat frequency that is more readily amplified.

high frequency. A frequency in the band extending from 3 to 30 MHz.

high-mu tube. A vacuum tube having a high amplification factor.

high-pass filter. A filter designed to pass currents at all frequencies above a desired frequency while attenuating the frequencies below the desired frequency.

high Q. Having a high ratio of reactance to effective resistance. Factor determining coil efficiency.

hookup wire. Usually tinned and insulated No. 18, 20, 22, or 24 soft-drawn copper wire. Used in wiring electronic circuits. May be solid or stranded.

horizontal blanking. The pulse which cuts off the electron beam while it is returning from the right side to the left side of the screen of a cathode-ray tube.

hum. A low and constant audio frequency, usually either 60 or 120 cycles, in the output of an audio amplifier. Hum is frequently caused by a faulty filter capacitor in the power supply or by heater-cathode leakage in a tube.

I$_p$. Symbol to designate plate current of a vacuum tube.

impedance. The total opposition that a circuit offers to the flow of alternating current or any other varying current at a particular frequency; a combination of resistance and reactance. The symbol for impedance is Z, the unit is the ohm.

impedance angle. Angle of the impedance vector with respect to the resistance vector, representing voltage lag or lead with respect to current.

impedance coil. A choke coil. An inductor.

impedance match. The condition in which the impedance of a component or circuit is equal to another impedance to which it is connected.

indirectly heated cathode. A cathode to which heat is supplied by an independent heater element in a thermionic tube.

induced. Produced as a result of the influence of an electric or magnetic field.

induced current. A current due to an induced voltage.

induced voltage. A voltage produced in a circuit by changes in the number of magnetic lines of force which are linking or cutting across the conductors of the circuit.

inductance. That property of a coil or other radio part which tends to prevent any change in alternating-current flow. Inductance is measured in henries.

inductance bridge. An instrument similar to a Wheatstone bridge, used to measure an unknown inductance by comparing it with a known inductance.

induction. The process by which an object is given an induced voltage by exposure to a magnetic field.

inductive circuit. Circuit containing for the most part inductive reactance, rather than capacitive reactance or simply pure resistance.

inductive coupling. A form of coupling in which energy is transferred from a coil in one circuit to a coil in another circuit by induction.

inductive feedback. Feedback of energy from the plate circuit of a vacuum tube to the grid circuit through an inductance or by means of inductive coupling.

inductive load. A load that is predominantly inductive. Also called lagging load.

inductive reactance. Reactance due to the inductance of a coil or other part in an alternating-current circuit. Inductive reactance is measured in ohms, and is equal to the inductance in henries multiplied by the frequency in cycles, times 2π.

inductor. A circuit component designed so that inductance is its most important property. Also a coil.

in phase. Condition existing when waves pass through maximum and minimum values of like polarity at the same instant.

input transformer. A transformer used to transfer incoming energy to the input of a circuit or device.

insulator. A device having high electric resistance, used for supporting or separating conductors so as to prevent undesired flow of current between conductors or to other objects.

interelectrode capacitance. The capacitance which exists between electrodes in a vacuum tube or a solid-state device.

interference filter. A device used between a source of interference and a radio, to attenuate or eliminate noise.

interlock. A safety device which automatically opens the AC supply circuit when an access door or cover to the circuit is opened.

intermediate frequency. In superheterodyne reception, a frequency resulting from the combination of the received frequency with the locally generated frequency.

intermediate-frequency amplifier. That section of a superheterodyne receiver which is designed to amplify signals at a predetermined frequency called the intermediate frequency of the receiver.

intermediate-frequency transformer. A transformer employed at the input and output of each intermediate-frequency amplifier stage in a superheterodyne receiver.

internal resistance. The resistance of a battery, generator, or circuit component.

International Morse Code. The code used universally for radio telegraphy.

interpolation. The process of finding the value between two known values.

interstage coupling. Coupling between vacuum-tube or transistor stages.

interstage transformer. A transformer used to provide coupling between two amplifier stages.

inverse feedback. See negative feedback.

ion. An atom or molecule which has fewer or more electrons than normal. A positive ion is one which has lost electrons, and a negative ion is one which has gained electrons.

ionization. The breaking up of a gas atom into two parts, a free electron and a positively charged ion.

ionization current. Current flow existing between two oppositely charged electrodes in an ionized gas.

ionization potential. The voltage required to ionize an atom or molecule.

ionosphere. The upper portion of the earth's atmosphere beginning at about 30 miles above the earth's surface.

IR drop. The voltage drop produced across a resistance R by the flow of current I through the resistor.

iron-core transformer. A transformer in which iron forms part or all of the magnetic circuit linking the windings.

isolation transformer. A transformer with independent primary and secondary windings. Transformers of this type (generally 1:1 ratio) are used to isolate AC-DC equipment from the AC power line.

jamming. Deliberate transmission of noise signals to interfere with reception of signals from another station.

jumper. A short length of conductor used to make a temporary connection.

K. Symbol for cathode or any numerical value that remains constant during a given period. Also, abbreviation for 1,000.

Kennelly-Heaviside layer. See Heaviside layer. This region above the earth is often called by both names because it was discovered almost at the same time by two scientists working independently.

key. A hand-operated switch used to send code signals by telegraphy or radiotelegraphy.

kilo-. Metric prefix meaning 1,000.

kilohertz. 1,000 hertz. Abbreviated kHz.

kilovolt. 1,000 volts. Abbreviated kv.

kilowatt. A unit of electrical power equal to 1,000 watts. Abbreviated kw.

Kirchhoff's current law. A fundamental law of electricity which states that the sum of all the currents flowing to a point in a circuit must be equal to the sum of all the currents flowing away from that point.

Kirchhoff's voltage law. A fundamental law of electricity which states that the sum of all the voltage sources acting in a complete circuit must be equal to the sum of all the voltage drops in that same circuit.

knife switch. A switch in which one or more flat metal blades, pivoted at one end, serve as the moving connectors, making contact with flat, gripping spring clips.

L. Symbol for coil or transformer winding.

lagging load. Inductive load; the current lags behind the voltage.

lambda. Greek letter λ, used to designate wavelength measured in meters.

laminated. A type of construction widely used for the cores of iron-core transformers, choke coils, and electromagnets. The desired shape of core is built with thin strips of a magnetic material such as soft iron or silicon steel.

L antenna. An antenna consisting of one or more horizontal wires with vertical lead-in connected at one end.

lattice-wound coil. A honeycomb coil. A coil wound so as to reduce distributed capacitance, having the appearance of lattice work.

layer winding. A coil-winding method in which adjacent turns are laid evenly side by side along the length of the coil.

LC **product.** Inductance *L* in henries, multiplied by capacitance *C* in farads.

LC **ratio.** Inductance in henries, divided by capacitance in farads.

lead-in. The conductor or conductors that connect the antenna proper to electronic equipment.

lead-in insulator. Generally, a tubular insulator inserted in a hole drilled through a barrier of some sort and through which the lead-in wire can be brought.

leakage current. Undesirable flow of current through or over the surface of an insulating material or insulator; or, the flow of direct current through a capacitor. Also, the alternating current that passes through a rectifier without being rectified.

leakage resistance. The resistance of the path over which leakage current flows, normally a high value.

left-hand rule. A rule for determining direction of magnetic lines of force around a single wire. If the fingers of the left hand are placed around the wire in such a way that the thumb points in the direction of *electron* flow, the fingers will then be pointed in the direction of the magnetic field.

LF, lf. Low frequency; the FCC designation for the band from 30 to 300 kc.

lightning arrester. A protective device which leaks off static charges in the vicinity of an antenna to ground, and thus tends to prevent the charges from building up to the intensity of lightning.

lightning rod. A metallic rod projecting above a structure, connected to ground.

linear. A relation such that any change in one of two related quantities is accompanied by an exactly proportional change in the other. In ham practice, the word linear denotes a relatively high-power amplifier added to a small transmitter to increase its signal output.

linear amplification. Amplification in which the wave form is reproduced accurately, but in magnified form.

linear modulation. Modulation which is equally proportional to the amplitude of the sound wave at all audio frequencies.

line cord. A two- or three-wire cord terminating in a two- or three-prong plug at one end and used to connect equipment to a power outlet.

line drop. The voltage drop between two points on a power line or transmission line.

line filter. A device inserted in the power line to block noise impulses which might otherwise enter the equipment from the power line.

line-voltage regulator. A device such as a ballast voltage regulator or special transformer that delivers an essentially constant voltage to the load, regardless of minor variations in the line voltage.

link coupling. Two or more coils of separate circuits coupled by a transmission line.

load line. A straight line drawn across a series of plate current plate voltage characteristic curves on a graph to show how plate current will change with grid voltage when a specified plate load resistance is used.

local oscillator. The oscillator of a superheterodyne receiver.

loctal tube. An eight-prong vacuum tube having a lock-in type of base.

long waves. Wavelengths longer than the longest broadcast-band wavelength of 545 meters. Long waves correspond to frequencies between about 30 and 550 kc.

loop antenna. An antenna consisting of one or more complete turns of wire. Loop antennas are also commonly used in direction-finding equipment and portable radios.

loopstick antenna. A built-in receiving antenna widely used in broadcast receivers. Loopstick antennas consist of a coil wound on a powdered-iron core. In some types the inductance is adjusted by moving the core.

loose coupling. A small amount of coupling between two coils or circuits.

loran. Long Range Navigation. A system used by ships and aircraft for fixing their own position from radio signals broadcast by two or more synchronized transmitting stations.

loudspeaker. A device for converting audio-frequency current into sound waves.

low frequency. A frequency in the band extending from 30 to 300 kc in the radio spectrum.

low-pass filter. A filter designed to pass currents at all frequencies below a critical frequency, while substantially attenuating the amplitude of other frequencies.

L pad. A dual volume control presenting a constant load impedance at all control settings.

lug. A small strip of metal placed on a terminal screw or riveted to an insulating material to provide a means for making soldered connections.

M. Abbreviation for mega, prefix meaning million. Commonly used as megohms, for resistors, and megahertz, for frequency figures. Also symbol for mutual inductance.

magnet. A metallic material which attracts iron and steel, and, if free to move, aligns itself north and south because of the influence of the earth's magnetic field.

magnetic deflection. Method of deflecting electrons in a cathode-ray tube by means of the magnetic field generally produced by coils placed outside the tube.

magnetic field. A region surrounding a magnet or a conductor through which current is flowing.

magnetic flux. The sum of all the magnetic lines of force from a magnetic source.

magnetic flux density. The number of magnetic lines of force per unit area.

magnetic focusing. A method of focusing an electron stream in a cathode-ray tube through the action of magnetic lens.

magnetic lines of force. Imaginary lines used to designate the directions in which magnetic forces are acting throughout the magnetic field associated with a permanent magnet, electromagnet, or current-carrying conductor.

magnetic poles. Regions of a magnet near which the field is concentrated, usually the two ends of a magnet. The north pole, the south pole.

magnetic shield. A soft-iron housing used to protect equipment or components from the effects of stray magnetic fields.

magnet wire. Insulated copper wire in sizes used for winding coils of electromagnetic devices.

master oscillator. An oscillator of comparatively low power used to establish the carrier frequency of a transmitter.

matching. Connecting two circuits or components with a coupling device so that the impedance of either circuit will be equal to the impedance existing between them.

matching transformer. See impedance-matching transformer.

mean carrier frequency. The center or resting frequency of a frequency-modulation transmission transmitter.

medium frequency. The band from 300 to 3,000 kHz.

meg. Widely used as abbreviation for megohm.

meg- or mega-. A prefix meaning one million times.

megahertz. One million hertz per second. Abbreviated MHz.

megohm. One million ohms. Abbreviated meg(s).

mercury battery. A type of battery especially characterized by extremely uniform output voltage and by very long shelf life. Mercury batteries use a zinc-powder anode; the cathode is mercuric oxide powder and graphite powder.

mercury switch. An electric switch made by placing a large globule of mercury in a glass tube with electrodes arranged so that tilting the tube will cause the mercury to make or break the circuit.

mercury-vapor rectifier. A rectifier to be containing mercury vapor. The gaseous discharge permits much larger anode currents than could be obtained in a high-vacuum tube of equivalent dimensions.

metal tube. A vacuum or gaseous tube having a metal envelope, with electrode leads passing through glass beads fused in the metal housing.

meter. A device that measures or registers an electric quantity. Also, the unit of length in the metric system (39.37 inches).

mho. The unit of conductance or admittance. It is the word ohm spelled backward.

mica. A transparent flaky mineral which splits into thin sheets and has excellent insulating and heat-resisting qualities. It is used to separate the plates of capacitors, to insulate electrode elements of vacuum tubes, and for other insulating purposes.

mica capacitor. A fixed capacitor employing mica as the dielectric.

micro-. A prefix meaning one-millionth of. Designated by the Greek letter μ (mu) in abbreviations.

microampere. One-millionth of an ampere. Also written μa.

microfarad. One-millionth of a farad. Correctly abbreviated as μf, but sometimes shown as uf, mf, or mfd.

microhenry. One-millionth of a henry. Also written μh.

micromicrofarad. One-millionth of a microfarad. Abbreviated $\mu\mu$f, uuf, mmf, or mmfd. Being replaced by a newer unit, picofarad, meaning the same thing.

microphone. A device which converts sound waves into corresponding audio-frequency electric energy. It contains some form of flexible diaphragm which moves in accordance with sound-wave variations. This movement, in turn, generates a minute voltage which is fed to an amplifier.

microphone preamplifier. An audio amplifier which initially amplifies the output of a microphone.

microphone transformer. An ironcore transformer used for coupling microphones to the audio amplifiers.

microswitch. Trade name for a small switch in which a minute motion makes or breaks contact.

microvolt. One-millionth of a volt.

microwaves. Electromagnetic waves whose frequencies are higher than 300 MHz.

mike. Colloquialism for microphone.

milli-. A prefix meaning one-thousandth of.

milliammeter. A meter calibrated in milliamperes.

milliampere. A unit of current equal to one-thousandth of an ampere. Abbreviated ma.

millihenry. A unit of inductance equal to one-thousandth of a henry. The plural is millihenries. Abbreviated mh.

millimeter. A metric unit of length equal to one-thousandth of a meter, or approximately $1/25$th inch (0.03937 inch). Abbreviated mm.

millivolt. A unit of voltage equal to one-thousandth of a volt. Abbreviated mv.

milliwatt. A unit of power equal to one-thousandth of a watt. Abbreviated mw.

mismatch. The conditions in which the impedance of a source does not match or equal the impedance of a connected load.

mixer. That stage in a superheterodyne receiver in which the incoming radio-frequency signal is combined with the signal from the local oscillator to produce the intermediate-frequency signal.

mixing. Combining two or more signals.

mobile receiver. A radio receiver designed to be operated while in motion, as in an automobile.

mobile transmitter. A radio transmitter designed to be operated while in motion.

modulate, modulation. To vary the amplitude, frequency, or phase of a radio-frequency carrier in accordance with speech, music, or other forms of signals.

modulation envelope. A curve drawn through the peaks of a graph showing the waveform of an amplitude-modulated signal.

modulator. An audio-frequency amplifier modulating a radio-frequency carrier signal.

monitor. A device used for checking radio or audio signals.

motorboating. Feedback occurring at a low audio-frequency rate in an audio amplifier. Resembles sounds made by a motorboat.

moving-coil loudspeaker. A loudspeaker in which a coil carrying the audio-frequency current is directly attached to the moving cone.

mu. Greek letter μ, a symbol for amplification factor and for the prefix micro-, one-millionth.

μa, ua. Microampere.

mu factor. The amplification factor of a tube.

multimeter. A test instrument for measuring voltage, current, and resistance. Volt-ohmmeters are of this type.

multiplier. A resistor used in series with a voltmeter or ohmmeter to increase the range of the meter.

mutual inductance. Between two coils, it is the flux linkage in either coil due to current flowing in the other.

μw. Microwatt.

mv. Millivolt.

mw. Milliwatt.

NC. No connection. Used on tube-base diagrams.

negative. A term used to describe a terminal from which electrons flow.

negative feedback. An arrangement by which a signal is fed back from the plate circuit to the grid circuit 180 degrees out of phase with the grid signal, thus decreasing gain. Also called inverse feedback.

neon. An inert gas used in some tubes, producing a bright orange-red glow when ionized. Neon-filled tubes are used as voltage regulators.

neutralization. The process of canceling the effects of interelectrode capacitance of an amplifier tube.

neutralizing capacitor. A capacitor, usually variable, employed in neutralizing circuits.

node. Any point in a wave system at which the amplitude is zero. The type of node is usually specified, since there can be nodes of voltage, current, etc.

noise. Interference characterized by undesirable random disturbances caused by internal circuit defects or from some external source. In radio receivers, noise appears as an audible hissing or crackling sound.

noise filter. A combination of one or more choke coils and capacitors used to block noise interference.

noise level. Volume of noise, usually expressed in decibels.

noise suppressor. A circuit used in a receiver or amplifier to reduce noise.

nonconductor. An insulating material.

noninductive load. A load having no inductance.

noninductive winding. A winding made so that one turn or section cancels the field of the next adjacent turn or section. For example, the wire may be doubled before winding. Used particularly with resistors to prevent them from exhibiting resonant effects.

nonlinear. Not directly proportional and hence producing a curve instead of a straight line when plotted graphically with linear coordinates.

nonlinear detection. Square law detection. Detection based on the curvature of a tube characteristic. This results in distortion without complete rectification. Either of these effects results in demodulation.

north pole. That pole of a magnet at which lines of force are considered as leaving; the lines enter the south pole.

nucleus. The central part of an atom. It consists of protons and neutrons, has a positive charge, and constitutes practically the entire mass of an atom.

null. Zero.

null indicator. Any device that indicates when current, voltage, or power is zero.

octal socket. A tube socket with openings for eight equally spaced pins, and a slot for aligning the center key.

ohm. It is that resistance across which 1 volt will cause a current of 1 ampere to flow.

ohmmeter. An instrument for measuring resistance.

Ohm's law. A fundamental law of electricity which expresses the relationship between voltage, current, and resistance in a DC circuit, or the relationship between voltage, current, and impedance in an AC circuit.

ohms-per-volt. A sensitivity rating for voltage-measuring instruments, obtained by dividing the resistance of the instrument in ohms at a particular range by the full-scale voltage value at that range. The higher the ohms-per-volt rating, the more sensitive is a meter.

omega. Greek letter ω used to represent the word ohm.

omnidirectional. In all directions, such as the radiation pattern of a vertical antenna.

open-circuit voltage. The voltage at the terminals of a voltage source when no current is flowing, i.e., with no load connected across the voltage source.

oscillation. Periodic variations in a system or circuit, especially those of alternating current.

oscillator. A circuit using vacuum tubes or transitors to generate alternating current over a wide range of frequencies.

oscillator coil. The transformer or coil used in an oscillator circuit.

oscilloscope. See cathode-ray oscilloscope.

output impedance. The impedance as measured between the output terminals of an electronic device, generally at a definite frequency or at a predominant frequency. For maximum power transfer, the load impedance should match or be equal to this output impedance.

output indicator. A meter or other device connected to indicate variations in signal strength of the output circuits.

output stage. The final stage of an electronic device.

output transformer. The iron-core audio-frequency transformer used to match the output stage of an audio-frequency amplifier to its loudspeaker or other load.

output tube. An amplifier tube used in an output stage.

overload. A load greater than a device is designed to handle.

overmodulation. Amplitude modulation in excess of 100 percent.

P. (1) Designation for the primary winding of a transformer. (2) Designation for the anode or plate of an electron tube.

padder. In a superheterodyne receiver, the capacitor placed in series with the oscillator tuning circuit to control the receiver calibration at the low-frequency end of a tuning range. Also, any small capacitor inserted in series with a main capacitor, for alignment purposes.

paper capacitor. A fixed capacitor consisting of strips of metal foil separated by an oiled or waxed paper dielectric.

parallel resonant circuit. A tuning circuit consisting of a coil and a capacitor connected in parallel. At resonant frequency it offers a high impedance.

parasitic oscillations. Unwanted self-sustaining oscillations at a frequency different from the operating frequency.

parasitic suppressor. A combination of inductance and resistance inserted in a circuit to suppress parasitic oscillations.

patch cord. A cord equipped with plugs at each end, used to connect two jack receptacles.

peak. The maximum instantaneous value of a quantity.

peak load. The maximum load consumed or produced in a given period of time.

peak plate current. The maximum instantaneous plate current flowing in a tube.

peaks. Momentary high amplitude levels occurring in electronic equipment.

peak voltmeter. A voltmeter that reads peak value of a voltage.

pentagrid converter. A tube employed as an oscillator-mixer in a superheterodyne receiver.

pentode. A vacuum tube having five electrodes.

period. The time required for one complete cycle of recurring quantity.

permanent-magnet loudspeaker. A moving-coil loudspeaker in which the magnetic field is produced by a permanent magnet.

permeability tuning. Tuning a resonant circuit by changing the coil inductance by positioning an iron core.

phase. The position at any instant which a periodic wave occupies in its cycle. If amplitude is plotted perpendicular to a time axis, phase may be represented as a position along the time axis. When the time of one period is 360 degrees, the phase position is called a phase angle.

phase difference. Relation between two sinusoidal quantities of the same frequency. It is the fraction of a cycle by which one of the waves would have to be moved along the time or frequency axis to make the two waves coincide. One quantity is considered as leading or lagging the other by the angle of the phase difference.

phenolic material. A thermosetting insulating plastic material used for countless electric, electronic, and mechanical applications.

Phillips screw. A screw with an indented "cross," instead of the conventional slot recessed in its head. Requires a Phillips screwdriver to remove or insert it.

phosphorescence. A form of light given off by a phosphor after the excitation light or electron stream has ceased. When emission of light occurs during excitation, the result is fluorescence.

photoelectric cell. A general term applying to any cell or tube whose properties are affected by illumination.

pickup.   A mechanical device that converts some form of intelligence into a corresponding electric signal. Also called transducer.

Pierce oscillator.   A crystal oscillator circuit featuring a crystal connected between the grid and plate of the oscillator tube.

piezoelectric.   Property of some crystals to generate a voltage when mechanical force is applied, and, conversely, the ability to produce a mechanical force by expanding or contracting whenever a voltage is applied.

pigtail.   A wire at each end of small components such as resistors and capacitors.

pilot lamp.   A small lamp used to illuminate the tuning dial of electronic equipment or as an indicator lamp.

pi network.   A network of three impedances, two across the line and the third inserted in one line between the other two, simulating the Greek letter π.

plate.   The common name for the principal anode of a vacuum tube. One of the conductive electrodes of a capacitor. Also, one of the electrodes of a storage battery.

plate bypass capacitor.   A capacitor connected to the plate circuit of a vacuum tube to bypass high-frequency currents.

plate circuit.   A circuit including the plate voltage source and all other parts connected between the cathode and plate terminals of a vacuum tube.

plate current.   The electron flow from the cathode to the plate inside a tube.

plate detection.   Detection of radio-frequency signals takes place in the plate circuit of a vacuum tube.

plate dissipation.   The amount of power lost as heat in the plate of a vacuum tube.

plate load impedance.   The impedance to current flow in the external circuit of a vacuum tube between its plate and cathode.

plate modulation.   The introduction of the modulating wave into the plate circuit of any tube in which the carrier-frequency wave is present.

plate resistance.   The ratio of a small change in plate voltage divided by a small change in plate current, in vacuum-tube circuits. The symbol is R_p.

plate supply.   The voltage source used in a vacuum-tube circuit to put the plate at a high positive potential with respect to the cathode.

plate voltage. The direct voltage between the plate and the cathode of a vacuum tube.

plug-in coil. A coil having as its terminals a number of prongs so that it may fit into a mounted socket.

PM. Permanent magnet.

polarity. An electric condition determining the direction in which current flows. Applied to DC sources and to components when connected in DC circuits.

polyethylene. A tough, flexible, plastic compound having excellent insulating properties.

polystyrene. A clear thermoplastic material having excellent dielectric properties.

porcelain. A glazed ceramic insulating material.

positive bias. The condition in which the control grid is positive with respect to the cathode of a vacuum tube.

positive feedback. See regeneration.

positive terminal. The terminal of a battery or other voltage source toward which electrons flow in the external circuit.

potential. Voltage.

potential difference. The difference in voltage at two points.

potentiometer. A variable resistor with connections to both ends of the resistance element and to the rotating arm that passes over the latter.

power. Rate of doing work. Energy per unit time. May be expressed in watts or in kilowatts (thousands of watts).

power amplification. A ratio of the power output of an amplifier to the power supplied to the input circuit.

power amplifier. An audio- or radio-frequency amplifier designed to deliver a relatively large amount of output energy. Also, the last stage of an amplifier as distinguished from previous stages usually classed as voltage amplifiers.

power gain. The ratio of two powers such as output to input of a vacuum tube or output to input of an audio-frequency amplifier.

power level. The amount of electric power passing through a given point in a circuit. Power level can be expressed in watts or in decibels.

power line. Two or more wires used for conducting power from one location to another.

power output. The power in watts delivered by an amplifier to a load, such as a speaker.

power pack. The power-supply unit of a radio receiver, amplifier, transmitter, or other radio apparatus.

power transformer. An iron-core transformer having a primary winding usually connected to an AC power line and having a number of secondary windings that provide different voltage values.

preamplifer. An extra stage of amplification at the input of an amplifier.

preselector A tuned-radio-frequency amplifier or antenna tuning device inserted between the receiver and the antenna to increase the amplitude of the incoming signal.

printed circuit. A method by which circuit connections and many of the components are printed or etched on a plane surface with conductive or resistive media for building compact circuits.

propagation. The travel of electromagnetic waves or sound waves through a medium.

protective gap. The space between two terminals across which transient voltages may arc, such as the gap in a lightning arrester.

push-pull circuit. A two-tube amplifier circuit in which the grid and plate of one tube are operating 180 degrees out of phase with the grid and plate of the other tube. Even-order harmonics are canceled. Push-pull circuits are used at both audio and radio frequencies.

push-pull oscillator. A vacuum-tube oscillator containing two tubes or a double-section tube connected in a phase relation similar to that of a push-pull amplifier.

push-pull transformer. An audio transformer designed for use in a push-pull amplifier circuit.

Q. A quality rating applied to a coil or resonant circuit. Q is the inductive reactance divided by the resistance.

QSL Card. A card exchanged by radio amateurs to confirm radio communication.

quarter-wave antenna. An antenna electrically equal to one-fourth the wavelength of the transmitted or received signal.

quartz crystal. A thin slice of quartz that vibrates at a frequency determined by its thickness and its original position in the natural quartz. Used to maintain high-frequency stability in oscillators.

quenching frequency. A locally generated frequency of a super-regenerative detector stage which prevents oscillation during reception of strong signals.

R.   (1) Letter used to denote resistance in ohms. (2) Symbol for resistor in a schematic diagram.

radar.   From the phrase, "Radio Detection and Ranging." Originally developed for wartime use, now widely used for such applications as marine and aeronautical navigation. It determines the presence and location of a distant object by transmitting high-power microwave pulses which are reflected back by the object to the radar unit. This reflected energy or "echo" appears as a "pip" on the screen of a cathode ray tube; the position of this pip on a calibrated time axis indicates the distance of the target from the radar unit. Position of the radar antenna indicates the bearing of the target in relation to the radar unit.

radiation.   Electromagnetic energy traveling outward into space such as radio waves, infrared rays, X rays, etc.

radiation pattern.   A diagram indicating the intensity of the radiation field of a transmitting antenna as a function of plane or solid angles. In the case of a receiving antenna, it is a diagram showing the response of the antenna to a unit field intensity signal arriving from different directions.

radio broadcasting.   A one-way transmission of voice and music.

radio channel.   A band of frequencies having sufficient width for radio communication and broadcasting purposes. The width of a channel depends on the type of transmission and the tolerance for the frequency of emission.

radio compass.   A radio direction finder with a zero center meter, used chiefly in marine and aircraft radio stations for navigational purposes.

radio converter.   A unit for adapting a receiver for use on frequency bands other than the ones for which it was designed originally.

radio direction finder.   A receiver and rotatable loop antenna used to determine the direction from which radio waves are being received.

radio frequency.   Specifically that part of the general frequency spectrum between audio sound and infrared light (about 20 kHz to 10,000,000 MHz). Generally, an AC frequency whose electromagnetic field can be radiated over great distances.

radio-frequency transformer.   A transformer for radio-frequency currents having either an air core or some form of pulverized-iron core.

radio spectrum.   The entire range of useful radio waves as classified into seven bands by the Federal Communications Commission.

Designation	Abbr.	Frequency	Wavelength
very low frequency	vlf	10–30 kHz	30,000–10,000 m
low frequency	lf	30–300 kHz	10,000–1,000 m
medium frequency	mf	300–3,000 kHz	1,000–100 m
high frequency	hf	3–30 MHz	100–10 m
very high frequency	vhf	30–300 MHz	10 to 1 m
ultrahigh frequency	uhf	300–3,000 MHz	100 to 10 cm
superhigh frequency	shf	3,000–30,000 MHz	10 to 1 cm

RC, RC **circuit.** Designation for any resistor-capacitor circuit.

RC **coupling.** Resistor-capacitor coupling between two circuits.

reactance. Opposition in ohms offered to the flow of alternating current by inductance or capacitance of a component or circuit.

reactive. Pertaining to either inductive or capacitive reactance.

rectification. The process of converting alternating current into a unidirectional current.

recitifier. A component that rectifies alternating current.

reflected wave. The sky radio wave, reflected back to the earth from an ionosphere layer.

refracted wave. The wave that is bent as it travels into a second medium, as from the atmosphere into an ionized layer of the stratosphere.

regeneration. A method of securing increased output from an RF amplifier by feeding part of the output back to the amplifier input so that it reinforces the input signal. Causes oscillation when carried to extremes.

regeneration control. A potentiometer or variable condenser which is used to control the amount of signal fed back from output to input in the regenerative detector stage.

regulated power supply. A power supply containing a regulator device for maintaining constant voltage or constant current under changing load conditions.

regulation. Holding constant some condition, like voltage, current, power, or position.

regulator. A device that accomplishes regulation within desired limits such as a current or voltage regulator.

relay. An electromagnetic switch employing an armature to open or close contactors.

relay rack. A standard vertical steel frame that accommodates standard-width (19-inch) panels of various heights on which are mounted electronic equipment.

remote cutoff tube. A variable mu tube. A tetrode or pentode in which the spacing of the control-grid wires is wider at the center than at the ends. Thus, the amplification of the tube does not vary in direct proportion to the bias, and some plate current flows regardless of the negative bias on the grid. Used in RF amplifiers.

resistance. The nonreactive opposition which a device or material offers to the flow of direct or alternating current. Resistance is measured in ohms, and is usually designated by the letter R.

resistance-capacitance-coupled amplifier. A vacuum-tube amplifier, the various stages of which employ resistors for the plate load, and in the grid circuit; coupling between them is by capacitors. Also used with transistors.

resistance-coupled amplifier. An amplifier in which the various stages are coupled solely by resistances between output and input. A direct-coupled amplifier.

resistance drop. Voltage drop due to flow of current through a resistance. also known as IR drop.

resistance wire. Wire made from an alloy having high resistivity.

resistor. A radio part which offers resistance to the flow of electric current. Its value is specified in ohms or megohms (1 megohm equals 1,000,000 ohms). A resistor also has a power-handling rating in watts, indicating the amount of power which can safely be dissipated as heat.

resonance. When reactance is zero or maximum in a circuit containing inductance and capacitance. If L and C are in series, circuit current is a maximum at resonance. If L and C are in parallel, external current supplied to circuit is a minimum at resonance and voltage nearly maximum.

resonance curve. A graphic representation showing the response of a resonant circuit to various frequencies within its operating range.

resonate. To bring to resonance, as by tuning.

response. Frequency range, or response, within specific limitations of speakers, amplifiers, etc.

return wire. A common wire, a ground wire, or the negative wire in a DC circuit.

RFC. Designation on diagrams to identify a radio-frequency choke coil.

rheostat. A resistor whose value can be changed smoothly.

rhombic antenna. A directional antenna array consisting of four long conductors laid out like an equal-sided parallelogram (rhombus).

ribbon microphone. A microphone with a moving conductor consisting of a single flexible ribbon of thin corrugated metal mounted between the poles of a permanent magnet. Also called velocity microphone.

rig. A system of components. An amateur rig is the complete amateur station consisting of receiver, transmitter, and all the accessory equipment.

ripple. The AC component present in the output of a DC generator, rectifier system, or power supply.

ripple current. The AC component of a pulsating unidirectional current.

ripple factor. Defined as the effective value of the alternating components of voltage (or current) divided by the direct or average values of the voltage (or current).

ripple filter. A low-pass filter designed to attenuate the AC components of a pulsating unidirectional current while passing the direct current from the rectifier or DC generator.

ripple frequency. The frequency of the ripple current.

ripple voltage. The alternating components of a unidirectional voltage.

rms. Root-mean-square.

root-mean-square. When referring to an AC value, the value that corresponds to the DC value that will produce the same heating effect. It is 0.707 of the peak AC value.

rosin-core solder. Solder made especially for electrical connections.

rotary beam antenna. A highly directional antenna that can be rotated by hand or by motor to any desired position. Provides maximum concentration of radiated energy or reception.

S. Letter used on circuit diagrams to denote a transformer secondary winding.

safety factor. The load, above the normal operating rating, to which a device can be subjected without failure.

saturation. The condition existing in a tube when tube current is the maximum that can be obtained by increasing the anode voltage. Also, the condition existing in a magnetic material when the flux density is the maximum that can be obtained by increasing the magnetomotive force.

scc wire. Single-cotton-covered wire.

sce wire. Single-cotton covering over enamel insulation on a wire.

schematic diagram. A diagram which shows the connections of an electronic device by means of symbols used to represent the parts.

screen. A metal partition or shield to isolate a device or apparatus from external magnetic or electric field. Also, the coated surface on the inside of the large end of a cathode-ray tube.

screen grid. A grid placed between the control grid and plate elements of a pentode or surrounding the plate of a tetrode. The purpose: to decrease grid-plate capacitance.

screen-grid modulation. A type of amplitude modulation wherein the modulating voltage is superimposed on the DC screen-grid voltage of the RF amplifer.

screen-grid voltage. The direct voltage applied between the screen grid and the cathode in a vacuum tube.

secondary. One or more transformer windings which receive energy by electromagnetic induction from a primary.

secondary voltage. The voltage across the secondary winding of a transformer.

second detector. In a superheterodyne receiver, the stage that separates the intelligence signal from the intermediate-frequency carrier signal.

selective. The characteristic of responding to a desired frequency to a greater degree than to other frequencies.

selective interference. Radio interference in a narrow band of frequencies.

selective reflection. Reflection of waves of only a certain group of frequencies.

selectivity. The ability of a radio receiver to reject undesired signals.

selenium rectifier. A dry-disk rectifier made of a crystalline selenium layer between two electrodes.

self-bias. Referring to a vacuum-tube stage which produces its own grid bias voltage.

self-healing capacitor. A capacitor that repairs itself after dielectric breakdown.

semiconductors. A class of semimetallic elements or oxides used in transistors, thermistors, and light-sensitive devices.

sensitivity. Characteristic of a radio or television receiver which determines the minimum input signal strength required for a given signal output value.

series resonant circuit. A circuit in which an inductor and a capacitor are connected in series, and have values such that the inductive reactance of the inductor is equal to the capacitive reactance of the capacitor at the resonant frequency. At resonance, the current though a series resonant circuit is a maximum.

SG. The screen-grid electrode of a vacuum tube.

sharp cutoff. Term applied to a tube or the grid of a tube in which the control grid spirals are uniformly spaced. The result is that as grid voltage is made negative, plate current decreases steadily to cutoff.

shielded line. A transmission line whose elements confine propagated radio waves inside a tubular conducting surface called the sheath. This prevents the line from radiating radio waves.

shielded pair. A two-wire transmission line surrounded by a metallic sheath.

shielded wire. Insulated wire covered with a metal shield, usually of tinned, braided copper wire.

short circuit. A low-resistance connection across a voltage source or a circuit, usually accidental, resulting in excessive current flow which often causes damage.

shorted out. Made inactive by connecting a heavy wire or other conductor path around a device or circuit, usually for protective purposes.

shortwaves. A general term usually applied to wavelengths whose frequency is higher than 1,600 kc.

shunt. Any part connected in parallel with some other part.

sidebands. Two bands of frequencies on either side of the carrier frequency of an amplitude-modulated carrier, including components whose frequencies are the sum and difference of the carrier and the modulation frequencies.

signal generator. An oscillator that generates audio- or radio-frequency signals at any frequency needed for aligning or servicing electronic equipment.

signal strength. A measure of the power output of a radio transmitter at a particular location. Usually expressed in microvolts or millivolts per meter of effective height of the receiving antenna employed.

signal-to-noise ratio. The ratio of the radio field intensity of a desired, received radio wave to the radio noise field intensity received with the signal.

sine wave. Wave form corresponding to a pure, single frequency oscillation. If amplitude is plotted against time, the curve is a sine function.

single-phase. Pertaining to a circuit or device that is energized by a single alternating voltage. One of the phases of a polyphase system.

single-sideband transmission. An important variation of amplitude-modulation transmission, wherein either of the two sidebands and the carrier wave are eliminated or greatly reduced. The remaining sideband carries the same intelligence as in the suppressed sideband, so normal communication is carried on. Commonly called SSB.

sinusoidal. Varying in proportion to the sine of an angle or time function. Ordinary alternating current is sinusoidal.

skin effect. The tendency of high-frequency currents to travel only on the outer surface of conductors.

skip zone. A region around a transmitter within which there is no reception from the transmitter.

sky wave. A radio wave that is reflected back to earth from the ionosphere. Sometimes called inosopheric wave.

slug. The movable iron core of an inductor; by moving the slug in or out the inductance is varied.

S meter. A signal-strength meter.

smoothing choke. An iron-core inductor employed as a filter to remove pulsations in the unidirectional output current of a rectifier.

smoothing filter. A filter composed of inductance and capacitance (or either alone) to remove AC components from the unidirectional output current of a rectifier or DC generator.

socket. A mounting device for tubes, plug-in coils, etc.

socket adapter. A device placed between a tube socket and a tube, to permit use of the tube in a socket designed for some other type of base, or to permit resistance or voltage measurements while the tube is in use.

solder. An alloy of lead and tin which melts at a fairly low temperature (about 500°F) and is used for making permanent electric connections between parts and wires.

solder gun. A soldering iron having an appearance similar to that of a pistol.

soldering iron. A device used to apply heat to a joint which is to be made permanent by soldering.

solid conductor. A single wire. A conductor that is not divided into strands.

sos. The international distress signal for radiotelegraphy.

sound. A vibration of a body at a rate which can be heard by human ears. The extreme limits of human hearing are about 20 hertz and 20,000 hertz. Sound can travel through any medium which possesses the ability to vibrate.

space charge. A gathering of electrons near the cathode of a vacuum tube. Being negative, it tends to limit the number of electrons which can reach the plate, for a given plate voltage.

space current. Current made up of electrons moving from the cathode to the plate in a vacuum tube.

spaghetti. Cloth or plastic tubing sometimes used to provide insulation over bare wires.

sparking. Intentional or accidental spark discharges, as between contacts of a relay or switch, or at any point at which an inductive circuit is broken.

sparkover. Ionization of the air between two electrodes permitting the passage of a spark.

speech amplifier. An audio-frequency amplifier used between a microphone and the input of the power amplifier to raise the output voltage of the microphone to the level required to guarantee the amplifier's full output.

splice. A connection of two or more conductors or cables to provide good mechanical strength as well as good conductivity.

spot. The luminous area produced on the viewing screen of a cathode-ray tube by the electron beam.

spreader. A short, light insulator used to hold apart the wires of an open transmission line.

spring-return switch. A switch which returns to its normal position when pressure is released.

SPST, spst. Single-pole, single-throw switch or relay.

spurious radiation. Any radiation from a radio transmitter at frequencies other than its operating frequency.

square wave. The wave form that shifts abruptly from one to the other of two definite values, giving a square or rectangular pattern when amplitude is plotted against time.

squealing. A condition in which a high-pitched note is heard along with the desired signal.

squelch circuit. An AVC circuit that reduces or attenuates the noise otherwise heard in a radio receiver between signals by blocking some stage when the signal amplitude is below a value called the squelch level.

stacked array. An array in which antenna elements are placed one above the other.

stage. All the components in a circuit containing one or more vacuum tubes or transistors performing a single function.

standard-frequency signal. Highly accurate signals broadcast by radio station WWV of the National Bureau of Standards in Boulder, Colorado. These signals are used throughout the world for calibration of radio equipment.

static. Noise heard in a radio receiver due to atmospheric electric disturbances such as lightning, or man-made causes such as electric motors, neon signs, or other appliances which produce sparking.

static charge. An electric charge accumulated on an object.

step-down transformer. A transformer in which the secondary delivers a lower voltage than is applied to the primary.

step-up transformer. A transformer in which the secondary delivers a higher voltage than is applied to the primary.

storage battery. A unit consisting of two or more storage cells.

storage cell. A voltaic cell which may be restored to a charged condition by an electric current opposite to that of the discharging current.

straight-line capacitance. A variable capacitor characteristic obtained when the rotor plates are shaped so that capacitance varies directly in proportion to the angle of rotation.

strand. One of the wires, or one of the groups of wires, of a multiwire conductor or cable.

stranded wire. A conductor composed of a group of wires or of any combination of groups of wires, usually twisted or braided together.

stray capacitance. Capacitance existing between circuit wires or parts, or between the metal chassis of electronic apparatus and the parts mounted on it.

stray field. Stray inductance. Leakage magnetic flux from an inductor.

superhet. Popular name for a superheterodyne receiver.

superheterodyne receiver. A type in which the incoming RF signals are sometimes amplified a small amount in the preselector, then fed into the frequency converter section (consisting of the oscillator, mixer, and first detector) for conversion into a fixed, lower carrier frequency called the intermediate frequency (IF) value of the receiver. The IF signals are highly amplified in the IF amplifier stages, then fed into the second detector for demodulation. The resulting audio signals are amplified in the conventional manner by the audio amplifier, then reproduced as sound waves by the loudspeaker.

superregenerative detector. A regenerative detector in which maximum regeneration is employed, but in which sustained oscillation is prevented by a separate quenching oscillator.

suppressor grid. A grid interposed between the screen grid and plate to prevent the passing of secondary electrons from the latter to the former.

suppressor modulation. A type of amplitude modulation in which the modulating voltage is superimposed on the suppressor grid.

surge. A sudden and transient variation in the current and/or voltage in a circuit.

sw. Abbreviation for switch. Used on diagrams.

s-w. Abbreviation for shortwave.

sweep circuit. A special oscillator circuit which generates a voltage having a sawtooth wave form for making the electron beam of a cathode-ray tube sweep back and forth across the fluorescent screen.

switch. A mechanical device for completing, interrupting, or changing the connections in an electric circuit.

sync. Abbreviation for synchronizing, usually voltage, as required in cathode-ray oscilloscopes.

T. Generally used to designate a transformer in circuit diagrams.

tank circuit. An inductor and a capacitor in a parallel-connected resonant circuit.

tap. A connection point or contact made in the body of a resistor or coil.

tapped resistor. A wire-wound fixed resistor having one or more taps. Also called an adjustable resistor.

telephony. Transmission and reproduction of audio sounds over connecting wires.

television. The transmission and reception of a rapid succession of images by means of radio waves traveling through space or over wires.

temperature-compensating capacitor. A capacitor whose capacitance varies with temperature.

terminated line. A transmission line terminated in the characteristic impedance of the line.

termination. The load connected to the output end of a transmission line.

tetrode. A four-electrode vacuum tube.

three-phase current. Current delivered through three or four wires, with the three current components differing in phase by ⅓ cycle or 120 electric degrees.

tight coupling. Closest possible coupling between two radio- or audio-frequency circuits.

time-delay relay. A relay in which the energizing or de-energizing of the coil precedes movement of the contact armature by a determinable interval.

tinned wire. Copper wire that has been coated with a layer of tin or solder during manufacture to simplify soldering.

tip. The contact at the end of a plug.

tip jack. A small single-hole jack into which a single-pin contact plug or tip is inserted to make a connection.

toggle switch. A small snap switch that is operated by a projecting lever.

tolerance. The permissible variation from rated or assigned value.

tone control. A device provided in electronic sound equipment to alter the proportion of bass and treble frequency response.

T-pad. A special type of potentiometer with equal input and output impedance. A T network.

transceiver. A combination transmitter-receiver in which a single set of tuning elements is used interchangeably for transmission and reception.

transconductance. The small change in plate current which results from a small change in grid voltage. Transconductance is equal to the amplification factor of a tube divided by the plate resistance.

transducer. Generally, a device which converts energy from one form into another, always retaining the characteristic amplitude variations of the energy being converted. Applied to both microphones and loudspeakers, more commonly to the latter.

transformer. An electric device that transfers electric energy by electromagnetic induction from one or more circuits to one or more other circuits. May be used to step voltage up or down. Transferred energy remains constant except for core and wire losses.

transformer oil. A high-quality insulating oil in which large power transformers are immersed for cooling and insulation, and to prevent oxidation.

transient oscillation. A momentary oscillation occurring in a circuit during switching.

transistor. A compact unit consisting of semiconducting material.

transmission. Transfer of electric energy from one location to another through conductors or by radiation or induction fields.

transmission line. A set of conductors used to transfer signal energy from one location to another, or to transmit current over long distances for power purposes.

transmission loss. A term used to denote a loss in power during the transmission of energy from one point to another.

transmitter. A term applying to the equipment used for generating an RF carrier signal, modulating this carrier with intelligence, and radiating the modulated RF carrier into space.

trap. Tuned circuit used to eliminate a given signal or to keep it out of a given circuit. A common trap is simply a tuned circuit which absorbs the energy of the signal to be eliminated.

trimmer capacitor. A small, adjustable capacitor, used in the tuning circuits of radio receivers and other radio apparatus.

triode. A three-electrode vacuum tube.

tubular capacitor. A paper or electrolytic capacitor having the form of a cylinder, with leads projecting axially from one or both ends.

tuned antenna. An antenna designed to provide resonance at the desired operating frequency by means of its own inductance and capacitance.

tuned circuit. An inductance-capacitance circuit that can be adjusted to resonance at a desired frequency.

tuned filter. An arrangement of electronic components tuned either to attenuate or pass signals at its resonant frequency.

tuned-grid tuned-plate oscillator. A vacuum-tube oscillator with tuned grid and plate circuits. Maximum oscillation depends on maximum feedback, which occurs when the grid and plate circuits are tuned to resonance.

tuned radio-frequency amplifier. An amplifier employing vacuum tubes or transistors and tuned circuits for the purpose of amplifying radio-frequency energy.

tuned radio-frequency stage. A stage of amplification which is tunable to the radio frequency of the signal being received.

tungsten filament. A filament used in incandescent lamps, in radio vacuum tubes, and in other tubes requiring an incandescent cathode.

tuning. Adjusting the inductance or capacitance (or both) in a coil-capacitor circuit. Or, adjusting circuits in electronic equipment for optimum performance.

tuning capacitor. A variable capacitor.

tuning coil. A variable inductor.

tuning meter. A DC meter connected to show when the receiver is accurately tuned to a desired frequency.

turns ratio. The ratio of the number of turns in a secondary winding of a transformer to the number of turns in the primary winding.

TVI. Television interference. Used by amateurs to denote interference by their transmitters with reception of picture or sound on television receivers.

twin line. A type of transmission line which has a solid insulating material, in which the two conductors are placed parallel to each other. Several impedance values are in common use (75, 150, and 300 ohms).

twisted pair. A cable composed of two insulated conductors twisted together either with or without a common covering.

ultrahigh frequency. A Federal Communications Commission designation for the frequency band from 300 MHz to 3,000 MHz.

unidirectional antenna. An antenna designed to radiate with maximum strength or receive with maximum sensitivity in a particular direction.

V. Letter used on diagrams to designate vacuum tubes.

va. Volt-ampere.

vacuum. An enclosed space from which practically all air has been removed.

vacuum capacitor. A type of capacitor having tubular elements which are housed in an evacuated glass envelope. Vacuum capacitors are characterized by extremely high breakdown voltage.

vacuum switch. A switch enclosed in an evacuated bulb.

vacuum tube. Specifically, an evacuated enclosure including two or more electrodes between which conduction through the vacuum may take place. A general term used for all electronic tubes.

vacuum-tube voltmeter. A test instrument which uses the high input impedance of a vacuum tube for measuring voltages without affecting the circuit being measured. Abbreviated VTVM.

valve. The term used in the British Commonwealth to designate a radio tube.

variable-mu tube. A remote cutoff tube. A vacuum tube with a grid designed so that the amplification factor and the mutual conductance are variable.

variable transformer. A transformer whose output voltage can be varied continuously.

velocity microphone. See ribbon microphone.

vernier. An auxiliary scale of slightly smaller divisions than the main measuring scale, permitting measurements with greater precision than allowed by the main scale.

vertical deflecting electrodes. The pair of electrodes that serves to move the electron beam up and down on the fluorescent screen of a cathode-ray tube employing electrostatic deflection.

vertically polarized wave. A wave whose direction of electric polarization is perpendicular to the earth.

vertical polarization. The condition in which radio waves are transmitted with their plane of electric polarization initially perpendicular to the surface of the earth.

vertical radiator. An antenna positioned perpendicular with respect to the earth and used for transmitting or receiving.

very high frequencies. A band of frequencies in the radio spectrum extending from 30 to 300 MHz. In television, Channels 2–13, or 54–216 MHz.

VHF. Very high frequency.

very low frequencies. A band of frequencies in the radio spectrum extending from 10 to 30 kHz.

vibration pickup. A microphone designed to respond to mechanical vibrations rather than to sound waves.

video amplifier. A stage in a television circuit which amplifies video frequencies.

vlf, VLF. Very low frequency.

voice coil. The moving coil that is attached to and drives the diaphragm or cone of a dynamic loudspeaker.

volt. The practical unit of voltage, potential, or electromotive force. One volt is the electromotive force which will move 1 ampere through a resistance of 1 ohm.

voltage. The electric pressure that makes current flow through a conductor. Same as electromotive force.

voltage divider. A resistor having one or more fixed or adjustable contacts along the length of its resistance element.

voltage doubler. A rectifier circuit that doubles the output voltage of a conventional rectifier.

voltage drop. The voltage developed by the flow of current through a resistance or impedance.

voltage feed. Excitation of a transmitting antenna by applying voltage at a voltage loop or antinode.

voltage gain. Voltage amplification.

voltage multiplier. A precision resistor used in series with a voltmeter to extend its measuring range.

voltage node. A point having zero voltage in a system of stationary waves.

voltage rating. The maximum sustained voltage that can safely be applied to or taken from a device without risking damage.

voltage regulation. The ability of a voltage source to maintain essentially constant output voltage in spite of variations in load.

voltage-regulator tube. A gas-filled electron tube used to keep voltage essentially constant despite wide variations in line voltage, or to maintain an essentially constant direct voltage in a circuit.

voltmeter. An instrument for measuring voltage.

volt-ohmmeter. A test instrument having provisions for measuring voltage, resistance, and current. Abbreviated VOM.

volume. The intensity or loudness of the sound produced by a headphone or loudspeaker.

volume control. A potentiometer used to vary the audio-frequency output of an audio amplifier.

VT. A symbol used on diagrams to indicate a vacuum tube.

W. Designates power in watts. The letter P sometimes is used alternatively with W.

walkie-talkie. A compact portable receiver-transmitter unit which is light enough to be carried in the hand or slung from the neck like a camera.

watt. The practical unit of electric power. In a DC circuit, equal to volts multiplied by amperes. In an AC circuit, true watts are equal to effective volts multiplied by effective amperes, then multiplied by the circuit power factor.

wattage rating. A rating expressing the maximum power which a device or component can safely absorb or handle.

wattmeter. A meter used to measure the power in watts or kilowatts which is being consumed by a device.

wave. A propagated periodic disturbance such as a radio, light, or sound wave.

wave band. A band of assigned frequencies.

wave form. The graphical representation of the shape of a wave, showing variations in amplitude versus time.

wavelength. The distance measured along the direction of propagation, between two points which are in phase on adjacent waves. A wavelength is the distance traveled by a wave in a time of one hertz.

wavemeter. A calibrated variable-frequency resonator used to determine wavelengths of radio waves.

wave trap. A device sometimes connected to the aerial system of a radio receiver to reduce the strength of signals at a particular frequency.

weak coupling. Loose coupling, in a transformer.

wet cell. A cell in which the electrolyte is in liquid form.

wet electrolyte capacitor. A capacitor employing a liquid electrolyte dielectric.

Wheatstone bridge. An instrument for measuring resistance. See bridge circuit.

winding. One or more turns of wire forming a continuous coil. Also, the coil itself, as in transformer windings.

wire gauge. In the American Wire Gauge system ("AWG") the relative sizes of wires are indicated by numbers from 1 through 40. No. 1, the largest in practical use, is 7.348 mm (close to 5/16 inch) in diameter; No. 40 is so thin it is scarcely visible to the naked eye.

wrinkle finish. A lacquer or varnish finish that shrinks and folds as it dries.

X cut. Term referring to the cut of a piezoelectric crystal which is made perpendicular to any two parallel faces.

Y cut. A quartz crystal cut such that the Y axis is perpendicular to the faces of the slab.

Z axis. The optical axis of a quartz crystal. It is perpendicular to both the X and Y axes. In cathode-ray oscilloscopes, variation of the beam intensity by an external voltage is called "Z-axis modulation."

zero adjuster. A device for bringing the pointer of an electric instrument or meter to zero when the electric quantity is zero.

zero beat.   The condition of a receiver in which an internal oscillator is at the exact frequency of an external radio wave so that no beat tone is produced or heard when the two are mixed.

zero bias.   A condition in which the control grid and cathode of a vacuum tube are at the same potential.

zero potential.   An expression usually applied to the potential of the earth, as a convenient reference for comparison.

Index

See the Glossary, pages 309-363, for additional terms and definitions.